Charles Gilbert Wheeler, Alfred Riche

Medical chemistry:

Including the outlines of organic and physiological chemistry: based in part upon

Riche's Manual de chimie

Charles Gilbert Wheeler, Alfred Riche

Medical chemistry:
Including the outlines of organic and physiological chemistry: based in part upon Riche's Manual de chimie

ISBN/EAN: 9783337732004

Printed in Europe, USA, Canada, Australia, Japan

Cover: Foto ©ninafisch / pixelio.de

More available books at **www.hansebooks.com**

MEDICAL CHEMISTRY,

INCLUDING THE OUTLINES OF

Organic & Physiological Chemistry.

BASED IN PART UPON RICHE'S MANUAL DE CHIMIE.

BY

C. GILBERT WHEELER,

Professor of Chemistry in the University of Chicago, and formerly
Professor of Organic Chemistry in the Chicago
Medical College.

SECOND AND REVISED EDITION.

PHILADELPHIA:
LINDSAY & BLAKISTON.

CHICAGO:
S. J. WHEELER.

1879.

CONTENTS.

	PAGE
INTRODUCTORY,	7
CLASSIFICATION OF ORGANIC COMPOUNDS,	10
HOMOLOGOUS SERIES,	12
HYDROCARBONS,	18
ALCOHOLS,	44
" MONATOMIC,	46
" DIATOMIC,	58
" TRIATOMIC,	64
ETHERS,	69
ALDEHYDS,	85
ACIDS,	90
" MONATOMIC,	96
" POLYATOMIC,	112
ALKALOIDS OR BASES,	127
" ARTIFICIAL,	132, 170
" NATURAL,	137
NEUTRAL FATTY BODIES,	174
SUGARS,	181
GLUCOSIDES,	193
VEGETABLE CHEMISTRY,	199
CELLULOSE,	205
STARCH,	210
DEXTRIN,	214
GUMS,	216

	PAGE.
ANIMAL CHEMISTRY,	221
ALBUMINOIDS,	225
FIBRIN,	231
CASEIN,	233
DIGESTION,	236
SALIVA,	237
GASTRIC JUICE,	242
BILE,	250
PANCREATIC JUICE,	261
CHYLE, LYMPH,	270
BLOOD,	272
HÆMOGLOBULIN,	285
CHEMICAL PATHOLOGY OF THE BLOOD,	294
RESPIRATION,	301
ANIMAL HEAT — MUSCULAR POWER,	316
ASSIMILATION,	321
SECRETION — THE URINE,	333
CHEMISTRY OF NORMAL URINE,	339
" " ABNORMAL "	347
URINARY SEDIMENTS,	352
" CALCULI,	353
ANALYSIS OF URINE,	356
" " URINARY DEPOSITS,	364
" " CALCULI,	368
SWEAT,	370
MILK,	376
THE SOFT TISSUES,	383
OSSEOUS TISSUE,	396
DENTAL "	403
EXUDATIONS,	407

PREFACE.

Medical chemistry has not as yet secured in American colleges sufficiently pronounced attention to create a demand for text-books of considerable size or extended scope. In these simple Outlines, therefore, no more has been attempted than this circumstance would appear to warrant. It is hoped that the necessary conciseness in method and form of expression has not resulted in any important sacrifice of perspicuity in thought or arrangement.

It would have been easier to prepare a larger work. From the bewildering wealth of results afforded by the labors of investigators in this branch of science, the appropriate selection of that suited to the wants of students was by no means an easy task.

It is assumed in these Outlines that those entering upon the study of Medical Chemistry have previously made themselves acquainted with Inorganic Chemistry as taught by some recent author, such as Miller or Barker, or have at least become familiar with the general principles of modern chemical philosophy. The author taking this for granted, has not, therefore, encumbered the work with a restatement of that which appertains to the theory of chemistry in general.

In addition to the organic portion of Riche's Manuel de Chimie, a translation of which by the author

PREFACE.

has served in part as basis for these Outlines, the works of Miller, Fownes, Williamson, Roscoe, and others have been freely used, while the chemical journals of Europe and America, including their latest numbers, have been consulted and the data which they afforded utilized.

Where the excerpta have been from journals of too recent issue to be found in standard authors, a reference in brackets has been made to the original source. Of the three series of numbers thus employed, the first has reference to the list of journals given at the close of this work, the second usually refers to the number of the volume, though sometimes to the year, the third indicates the page.

Lest any regard the number of characteristic reactions of the more important compounds as insufficient, it should be stated, that it was not within the plan of the author to adapt this work to the requirements of an analytical manual. Not more than two or three analytical tests are therefore given as a rule, and even this number only in the case of the leading compounds. A similar explanation might be proffered to any who may miss the full technical details relative to certain compounds which are usually given in works on applied, or technological chemistry.

Throughout the work, the centigrade thermometer and the metric system of weights and measures are employed, unless otherwise specifically stated.

C. GILBERT WHEELER.

UNIVERSITY OF CHICAGO, December, 1878.

ORGANIC CHEMISTRY.

INTRODUCTORY.

Organic chemistry is the science of the compounds of carbon.

Only a small number of other elements are met with in natural organic substances; they are hydrogen, oxygen and nitrogen, sometimes also, sulphur, phosphorus, and very rarely certain other elements.

Chemists have succeeded in incorporating most of the elemental substances in organic bodies, yet the larger number even of the artificial compounds include only the four elements first named.

Paraffine is found by analysis to contain only carbon and hydrogen, and is therefore called a *hydrogen-carbide*. The hydrocarbides are compounds so stable and fundamental that some chemists, as Schorlemmer for instance, have even defined organic chemistry as "the chemistry of hydrocarbons and their derivatives."

From alcohol, or sugar, we may obtain carbon and water. These bodies therefore are composed of three elements: carbon, hydrogen and oxygen, and are called *carbohydrates*; though by some chemists, this term is restricted to those compounds containing carbon with hydrogen, and oxygen in such proportions as would form water.

If albumen is decomposed by heat, the result is not only carbon and water, but also ammonia; this substance accordingly is *nitrogenous*.

The number of organic bodies is very great. As they are composed of a small number of elements only, it may be concluded that the latter unite in a very great variety of proportions; it is therefore of much importance to know the molecular grouping of these elements. The mere fact that the kind and number of elements entering into a compound are known, is not sufficient proof that its molecular structure is really determined. Synthesis must often be employed to confirm the results of analysis.

Berthelot has specially occupied himself with the synthesis of organic bodies, and has artificially produced a great number of them. Other chemists have experimented in the same direction during the last 15 or 20 years. However, Gerhardt's opinion advanced in 1854; viz., "The vital force alone operates by synthesis and reconstructs the edifice demolished by chemical affinity," has ceased to be held as true.

ISOMERISM.

Carbon, hydrogen, oxygen and nitrogen are not only capable of uniting in a great variety of proportions, but these elements also furnish numerous *isomeric* bodies; these comprise substances which, while com-

ISOMERISM. 9

posed of the same elements, have different properties. Sometimes the physical properties alone are different; we then have *physical isomerism*.

When the chemical properties themselves are modified, this is denominated *chemical* isomerism. Of the latter, two kinds are recognized.

I. *Polymerism;* cyanogen and paracyanogen are examples of this variety of isomerism; the latter is to be considered as cyanogen, CN condensed, thus $(CN)n$; it is a polymeride of cyanogen. The weight of the molecule of these two substances is therefore different.

II. *Metamerism.* At other times the isomerism results from a different grouping of elements in the compound, the molecular weight remaining the same.

We will illustrate this by two examples:
    a) Methyl acetate,
    and b) Ethyl formiate.
Acetic acid $= H\text{-}O\text{-}C_2H_3O$.
Methyl hydrate, or methyl alcohol $= H\text{-}O\text{-}CH_3$.
When these two bodies react they furnish water and methyl acetate, $CH_3\text{-}O\text{-}C_2H_3O = C_3H_6O_2$.
Formic acid $= H\text{-}O\text{-}CHO$.
Ethyl hydrate, or ethyl alcohol $= H\text{-}O\text{-}C_2H_5$.
Now formic acid contains CH_2 less than acetic acid, and hydrate of ethyl contains one molecule of CH_2 more than does hydrate of methyl. As these substances in reacting lose one molecule of water, it is therefore clear that the compound obtained will have, like the preceding one, the formula $C_3H_6O_2$. But these

two products are not identical substances, for the former treated with alkalies regains the molecule of water which it had lost, reforming acetic acid and methyl hydrate, while the latter regenerates formic acid and ethyl hydrate.

These bodies accordingly differ in the arrangement of their molecule; they are called *metameric bodies*.

Finally there exist bodies which are *isomeric, properly so-called*, possessing the same formula, having the same general reactions, the same chemical functions, and which differ only in a very few, chiefly physical, properties: such are oil of turpentine and oil of lemon, each having the formula $C_{10}H_{16}$.

CLASSIFICATION OF ORGANIC COMPOUNDS.

CHEMICAL TYPES.—The idea of referring organic bodies to some simple *model*, or *type*, was originally worked out by Laurent and Gerhardt, 1846-53, though the germs of their ideas on classification are to be found in the earlier papers of the distinguished American chemist T. Sterry Hunt. (*Am. Jour. Sci.* [2] xxxi.)

The four principal types are:

I. The hydrogen type, $\left.\begin{array}{l}H'\\H'\end{array}\right\}$ or H_2.

II. The oxide or water type, $\left.\begin{array}{l}H'\\H'\end{array}\right\} O''$ or H_2O.

III. The nitride or ammonia type, $\left.\begin{array}{l}H'\\H'\\H'\end{array}\right\} N'''$ or H_3N.

ORGANIC TYPES.

IV. The marsh gas type $\left.\begin{array}{l}H'\\H'\\H'\\H'\end{array}\right\} C^{IV}$ or H_4C.

Of the leading groups of organic bodies, we refer to the hydrogen type: hydrocarbides, aldehyds and the compounds of metals and metalloids with organic radicals.

To the water type are referred the alcohols, ethers, mercaptans and anhydrides.

To the ammonia type belong the amides, amines, and alkalamides, all of which are denominated *compound ammonias*.

Marsh-gas is the type to which carbon dioxide is referred, as well as some of the more complex organo-metallic bodies.

Further details as to the relation of each of these classes of compounds to their respective types will be given as each particular class is studied.

Besides the simple type, Kekulé has proposed compound types formed by the combination of two of the four types already given. Thus the types of ammonia and water combined serve as a pattern for carbamic and oxamic acids:

Carbamic acid. Oxamic Acid.

$\left.\begin{array}{l}H'\\H'\\H'\end{array}\right\} N'''$; $\left.\begin{array}{l}H\\H\\CO''\\H\end{array}\right\} N$, $\left.\begin{array}{l}H\\H\\(C_2O_2)''\\H\end{array}\right\} N$.
$\left.\begin{array}{l}H'\\H'\end{array}\right\} O'$ $\left.\begin{array}{l}\\ \\H\end{array}\right\} O$ $\left.\begin{array}{l}\\ \\H\end{array}\right\} O$

ORGANIC CHEMISTRY.

HOMOLOGOUS SERIES.

The members of a series of compounds which have the common difference of CH_2 are said to be *homologous*. Two or more such homologous series are termed *isologous*.

The first idea of progressive series in organic chemistry was enunciated by James Schiel, of St. Louis, Mo., in 1842. It was afterwards adopted by Gerhardt unchanged, save only in name. (100-5-195.)

The subjoined table will illustrate the nature of these series. Each vertical column forms a homologous series in which the terms differ by CH_2, and each horizontal line an isologous series in which the successive terms differ by H_2. The bodies of these last series are designated as the monocarbon, dicarbon group, etc.

$C H_4$ $C H_2$
C_2H_6 C_2H_4 C_2H_2
C_3H_8 C_3H_6 C_3H_4 C_3H_2
C_4H_{10} C_4H_8 C_4H_6 C_4H_4 C_4H_2
C_5H_{12} C_5H_{10} C_5H_8 C_5H_6 C_5H_4 C_5H_2
C_6H_{14} C_6H_{12} C_6H_{10} C_6H_8 C_6H_6 C_6H_4 C_6H_2.

The terms of the same homologous series resemble one another in many respects, exhibiting similar transformations under the action of given re-agents, and a regular gradation of properties from the lowest to the highest; thus, of the hydro-carbons, $C_n H_{2n+2}$, the lowest terms CH_4, C_2H_6, and C_3H_8, are gaseous at ordinary temperatures, the highest containing 20 or more carbon-atoms, are solid, while the intermediate compounds are liquids, becoming more and more viscid and less volatile, as they contain a greater number of carbon-atoms, and exhibiting a constant rise of about 20° C. (36° F.) in their boiling points for each addition of CH_2 to the molecule.

The individual series are given in the following table, with the names proposed for them by A. W. Hoffmann:

Methane	Methene			
CH_4	CH_2			
Ethane	Ethene	Ethine		
C_2H_6	C_2H_4	C_2H_2		
Propane	Propene	Propine	Propone	
C_3H_8	C_3H_6	C_3H_4	C_3H_2	
Quartane	Quartene	Quartine	Quartone	Quartune
C_4H_{10}	C_4H_8	C_4H_6	C_4H_4	C_4H_2
Quintane	Quintene	Quintine	Quintone	Quintune
C_5H_{12}	C_5H_{10}	C_5H_8	C_5H_6	C_5H_4
Sextane	Sextene	Sextine	Sextone	Sextune
C_6H_{14}	C_6H_{12}	C_6H_{10}	C_6H_8	C_6H_6

The formulæ in the preceding tables represent hydrocarbons all of which are capable of existing in the separate state, and many of which have been actually obtained. They are all derived from saturated molecules, C_nH_{2n+2} by abstraction of one or more *pairs* of hydrogen-atoms.

But a saturated hydrocarbon, CH_4, for example, may

give up 1, 2, 3, or any number of hydrogen-atoms in exchange for other elements; thus marsh gas, CH_4, subjected to the action of chlorine under various circumstances, yields the substitution-products,

$$CH_3Cl, \quad CH_2Cl_2, \quad CHCl_3, \quad CCl_4,$$

which may be regarded as compounds of chlorine with the radicles,

$$(CH_3)', \quad (CH_2)'', \quad (CH)''', \quad C^{iv};$$

and in like manner each hydrocarbon of the series, C_nH_{2n+2}, may yield a series of radicles of the forms,

$$(C_nH_{2n+1})', \quad (C_nH_{2n})'', \quad (C_nH_{2n-1})''' \quad (C_nH_{2n-2})^{iv}, \&c.$$

each of which has an equivalent value, or combining power, corresponding with the number of hydrogen-atoms abstracted from the original hydrocarbon. Those of even equivalence contain even numbers of hydrogen-atoms, and are identical in composition with those in the table above given; but those of uneven equivalence contain odd numbers of hydrogen-atoms, and are incapable of existing in the separate state, except, perhaps, as double molecules.

These hydrocarbon radicles of uneven equivalence are designated by Hoffmann, with names ending in *yl*, those of the univalent radicles being formed from methane, ethane, &c., by changing the termination *ane* into *yl ;* those of the trivalent radicles by changing the final *e* in the names of the bivalent radicles, methene, &c., into *yl;* and similarly for the rest. The names of the whole series will therefore be as follows :

CH_4	$(CH_3)'$	$(CH_2)''$	$(CH)'''$
Methane	Methyl	Methene	Methenyl
C_2H_6	$(C_2H_5)'$	$(C_2H_4)''$	$(C_2H_3)'''$
Ethane	Ethyl	Ethene	Ethenyl
C_3H_8	$(C_3H_7)'$	$(C_3H_6)''$	$(C_3H_5)'''$
Propane	Propyl	Propene	Propenyl
&c.		&c.	&c.

From these hydrocarbon radicles, others of the same degree of equivalence may be derived by partial or total replacement of the hydrogen by other elements, or compound radicles. Thus from propyl, C_3H_7, may be derived the following univalent radicles:—

C_3H_6Cl	$C_3H_3Cl_4$	C_3H_5O
Chloropropyl	Tetrachloropropyl	Oxypropyl
$C_3H_2Cl_3O$	$C_3H_6(CN)'$	$C_3H_6(NO_2)$
Trichloroxypropyl	Cyanopropyl.	Nitropropyl
$C_3H_4(NH_2)O$	$C_3H_6(CH_3)$	$C_3H_5(C_2H_5)_2$
Amidoxypropyl	Methylpropyl	Diethylpropyl.

From the radicles above mentioned, all well-defined organic compounds may be supposed to be formed by combination and substitution, each radicle entering into combination, just like an elementary body of the same degree of equivalence.

ORGANIC CHEMISTRY.

TABLE TO ILLUSTRATE THE ARRANGEMENT OF THE MORE

Series.	Hydro-carbons.	Sulphides.	Chlorides or Haloid Ethers.	Alcohols.
General Formula.	C_nH_{2n}	$\left. \begin{array}{l} C_nH_{2n+1} \\ C_nH_{2n+1} \end{array} \right\} S$	$C_nH_{2n+1}Cl$	$\left. \begin{array}{l} C_nH_{2n+1} \\ H \end{array} \right\} O$
1.	$C\ H_2$	$(CH_3)_2S$	$CH_3\ Cl$	$CH_3\ HO$
2.	$C_2\ H_4$	$(C_2H_5)_2S$	$C_2H_5\ Cl$	$C_2H_5\ HO$
3.	$C_3\ H_6$		$C_3H_7\ Cl$	$C_3H_7\ HO$
4.	$C_4\ H_8$		$C_4H_9\ Cl$	$C_4H_9\ HO$
5.	$C_5\ H_{10}$	$(C_5H_{11})_2S$	$C_5H_{11}Cl$	$C_5H_{11}HO$
6.	$C_6\ H_{12}$			$C_6H_{13}HO$
7.	$C_7\ H_{14}$			
8.	$C_8\ H_{16}$		$C_8H_{17}Cl$	$C_8H_{17}HO$
9.	$C_9\ H_{18}$			
10.	$C_{10}H_{20}$			
Types......	$\left. \begin{array}{l} H \\ H \end{array} \right\{$	$\left. \begin{array}{l} H \\ H \end{array} \right\} O$	$\left. \begin{array}{l} H \\ H \end{array} \right\{$	$\left. \begin{array}{l} H \\ H \end{array} \right\} O$

ORGANIC COMPOUNDS.   17

IMPORTANT ORGANIC COMPOUNDS IN HOMOLOGOUS SERIES.

Mercaptans.	Aldehyds.	Acids.	Simple Ethers.	Compound Ethers.	
$\begin{matrix}C_nH_{2n+1}\\H\end{matrix}\Big\}S$	$\begin{matrix}C_nH_{2n-1}\,O\\H\end{matrix}\Big\}$	$\begin{matrix}C_nH_{2n-1}O\\H\end{matrix}\Big\}O$	$\begin{matrix}C_nH_{2n+1}\\C_nH_{2n+1}\end{matrix}\Big\}O$	$\begin{matrix}C_nH_{2n+1}\\C_nH_{2n-1}O\end{matrix}\Big\}O$	
$CH_3\ HS$	$C\ H\ O,H$	$HC\ H\ O_2$	$(CH_3)_2O$	$C_2H_5\ C\ H\ O_2$	1.
$C_2H_5\ HS$	$C_2\ H_3\ O,H$	$HC_2\ H_3\ O_2$	$(C_2H_5)_2O$	$C_2H_5\ C_2\ H_3\ O_2$	2.
	$C_3\ H_5\ O,H$	$HC_3\ H_5\ O_2$		$C_2H_5\ C_3\ H_5\ O_2$	3.
$C_4H_9\ HS$	$C_4\ H_7\ O,H$	$HC_4\ H_7\ O_2$		$C_2H_5\ C_4\ H_7\ O_2$	4.
$C_5H_{11}HS$	$C_5\ H_9\ O,H$	$HC_5\ H_9\ O_2$	$(C_5H_{11})_2O$	$C_5H_{11}C_5\ H_9\ O_2$	5.
	$C_6\ H_{11}O,H$	$HC_6\ H_{11}O_2$		$C_2H_5\ C_6\ H_{11}O_2$	6.
	$C_7\ H_{13}O,H$	$HC_7\ H_{13}O_2$		$C_2H_5\ C_7\ H_{13}O_2$	7.
		$HC_8\ H_{15}O_2$		$C_2H_5\ C_8\ H_{15}O_2$	8.
		$HC_9\ H_{17}O_2$		$C_2H_5\ C_9\ H_{17}O_2$	9.
	$C_{10}H_{19}H,O$	$HC_{10}H_{19}O_2$	$(C_{10}H_{21})_2O$	$C_2H_5\ C_{10}H_{19}O_2$	10.
$\begin{matrix}H\\H\end{matrix}\Big\}O$	$\begin{matrix}H\\H\end{matrix}\Big\}$	$\begin{matrix}H\\H\end{matrix}\Big\}O$	$\begin{matrix}H\\H\end{matrix}\Big\}O$	$\begin{matrix}H\\H\end{matrix}\Big\}O$	

HYDROCARBONS.

The origin or preparation of these compounds, also called *hydrocarbides*, and their properties, physical and chemical, all differ largely.

They are unlike the hydrogen combinations studied in inorganic chemistry inasmuch as they possess but feeble chemical energy. Among the carbides are: acetylene, marsh-gas or methane, ethylene, oil of turpentine and of lemon, benzol, naphthalin, petroleum, caoutchouc, gutta-percha, etc.

The hydrocarbides will be divided into six series, they are all built upon the type of a molecule of hydrogen, or $\left.\begin{array}{l}H' \\ H'\end{array}\right\}$.

FIRST SERIES.

General Formula, $C_n H_{2n-2}$.

ACETYLENE, OR DIHYDROGEN DICARBIDE.

Discovered by Davy and composition determined by Berthelot.
Formula, C_2H_2.
Specific Gravity, 0.92. Density, 13. Molecular weight, 26.

Direct combination of Carbon and Hydrogen.
Up to comparatively recent times it has been considered impossible to unite carbon and hydrogen directly. Berthelot, however, succeeded in doing this in the year 1863.

PREPARATION.—The apparatus which he employed

in this remarkable synthesis, consisted of a glass flask, provided with two lateral tubulures through which passed two metallic rods, terminating in carbon points, and which approached so as to form, when connected with a powerful battery, an electric arc. The corks through which these rods passed were provided with another opening each, to which a tube was adapted. Through one of these tubes hydrogen was admitted and through the other the products of the reaction passed as they were formed.

The gas was collected in a solution of cuprous chloride in ammonia. A red-precipitate, acetylide of copper was formed, which was thrown upon a filter and treated with hydrochloric acid in a flask, whereupon acetylene was set free.

Many organic compounds produce acetylene on subjecting their vapors to the action of electric discharges.

Acetylene is also produced, as a rule, whenever organic matter is decomposed by heat.

PROPERTIES.—Acetylene is a colorless gas, having a disagreeable odor. It is moderately soluble in water, and is difficultly liquified. It is decomposed, at about the temperature at which glass melts, into carbon, hydrogen, ethylene, ethyl hydride and condensed hydrocarbides, among which Berthelot has found benzol. Thenard has recently obtained it both as a liquid and a vitreous solid. (9—78—219.)

Acetylene burns with a fuliginous flame. It detonates violently and without residue when mixed with

2.5 volumes of oxygen. Cuprous acetylide is an explosive body. It is sometimes formed in brass gas-pipes, and has been the cause of fatal accidents.

Chlorine acts upon acetylene with extreme energy; there is often detonation accompanied by light. On moderating the action the compound $C_2H_2Cl_2$ can be obtained, which, as well as the body $C_2H_2Cl_4$, can also be prepared by the action of antimonic chloride upon acetylene.

As acetylene is not uncommonly studied in connection with inorganic compounds, a more detailed account of this hydrocarbide need not be given here.

Acetylene is the prototype of a homologous series of hydrocarbides, of which the general formula is,

$$C_n H_{2n-2}.$$

The following members of this series are known:

Allylene,	$C_3 H_4$
Crotonylene,	$C_4 H_6$
Valerylene,	$C_5 H_8$
Rutylene,	$C_{10}H_{18}$
Benzylene,	$C_{15}H_{28}$.

ETHYLENE. 21

SECOND SERIES.

General formula, C^nH_{2n}.

ETHYLENE.

Synonyms: Elayl, Olefiant gas.
Formula C_2H_4.
Sp. Gr. 0.97. Molecular weight, 28.

This gas, for no good reason other than custom, is always studied in inorganic chemistry, usually in connection with the consideration of illuminating gas, of which, with methane, it forms a prominent constituent.

Ethylene is the type of a class of homologous hydrocarbides, of which the general formula is:

$$C_n H_{2n}.$$

Each member of the series is related to an alcohol from which it may be obtained on treatment with bodies having a great affinity for water, as sulphuric acid or zinc chloride.

$$C_n H_{2n+2} + O_2 - H_2O = C_nH_{2n}$$

We note the following members of this series:

Ethylene,	C_2H_4
Propylene,	C_3H_6
Butylene,	C_4H_8
Amylene,	C_5H_{10}
Hexylene,	C_6H_{12}
Heptylene,	C_7H_{14}
Octylene,	C_8H_{16}
Nonylene,	C_9H_{18}
Paramylene,	$C_{10}H_{20}$
Cetene,	$C_{11}H_{22}$
Duodecylene,	$C_{12}H_{24}$
Tridecylene, (Paraffin?)*	$C_{13}H_{26}$
Tetradecylene,	$C_{14}H_{28}$.

*A. G. Pouchet (66—[3] 4—868) has prepared from paraffin, by oxydation with nitric acid, paraffin acid, $C_{24}H_{43}O_2$, from which he deduces $C_{24}H_{50}$ as the formula for paraffin.

THIRD SERIES.

General formula, $C_n H_{2n+2}$

METHANE.

Discovered by Volta in 1778.
Synonyms; Methyl hydride, Marsh gas, Formene.
Formula CH_4 or CH_3, H.
Sp. Gr. 0.559. Molecular weight, 16.
Permanent gas, not liquifiable, neutral.

Not discussed in detail here for the same reasons as given under Ethylene.

Methane is the first member of the following very important homologous series:

$C H_4$	methyl hydride,	or	methane.
$C_2 H_6$	ethyl	"	" ethane.
$C_3 H_8$	propyl	"	" propane.
$C_4 H_{10}$	butyl	"	" butane.
$C_5 H_{12}$	amyl	"	" amane.
$C_6 H_{14}$	hexyl	"	" hexane.
$C_7 H_{16}$	heptyl	"	" heptane.
$C_8 H_{18}$	octyl	"	" octane.
$C_9 H_{20}$	nonyl	"	" nonane.
$C_{10} H_{22}$	decyl	"	" decane.
$C_{11} H_{24}$	undecyl	"	" undecane.
$C_{12} H_{26}$	bidecyl	"	" bidecane.

$C_{13}H_{28}$ tridecyl " " tridecane.
$C_{14}H_{30}$ tetradecyl " " tetradecane.
$C_{15}H_{32}$ pentadecyl " " pentadecane.
$C_{16}H_{34}$ hexadecyl " " hexadecane.

Nearly all the members of this series have been found in American petroleum, mixed with members of the preceding, or ethylene, series.

Crude petroleum, refined by fractional distillation, is still a mixture of various hydrocarbons.

The commercial names given to the products separated at the different boiling points, do not appertain to chemical compounds, or bodies having a definite composition.

Subjoined is a table based on Dr. C. F. Chandler's Report on Petroleum, (100—'72–41) showing the

PRODUCTS OF THE DISTILLATION OF CRUDE PETROLEUM.

NAME.	PERCENTAGE YIELDED.	SPECIFIC GRAVITY.	BOILING POINT.	CHIEF USES.
Cymogene			0°C.	Generally uncondensed—used in ice machines.
Rhigolene		.625	18.3	Condensed by ice and salt—used as an anæsthetic.
Gasolene	1½	.665	48.8	Used in making "air-gas."
C Naphtha	)	.706	82.2	Used for oil-cloths, cleaning, adulterating kerosene, etc. For paints and varnishes.
B Naphtha	-10	.721	104.4	
A Naphtha	)	.742	148.8	
Benzine	4			Used to adulterate kerosene oil.
Kerosene oil	55	.804	176.6	Ordinary oil for lamps.
Mineral sperm		.847	218.3	
Lubricating oil		.833	301.6	Lubricating machinery.
Paraffin	19½	Solid.		Manufacture of candles.

*Re-arranged from Dr. C. F. Chandler's Report on Petroleum, presented to the Board of Health, of the City of New York, 1870.

UNSAFE KEROSENE.

Many accidents occur by explosion of lamps, when kerosene oil contains too much of the lighter oils, benzine and naphtha. This makes the oil too readily inflammable, for the lighter oils are driven out by heating (as when a lamp or kerosene stove is burning), and their vapors mixed with the oxygen of the air form a dangerous explosive mixture. There is a law requiring manufacturers to keep kerosene oil free from these lighter oils, unfortunately not always faithfully enforced.

The temperature at which kerosene, on heating in an open vessel, emits vapors which readily catch fire on approaching a burning body, is called, technically, the "flash point," and that at which the kerosene itself inflames is called the "burning point."

FOSSIL RESINS, AND BITUMEN.

These substances include amber, retinasphalt, asphalt, retinite, and many other allied bodies which are chiefly contained in the tertiary strata. In many instances they are the products of the action of an elevated temperature upon vegetable bodies; and when this is the case, they form irregular deposits which impregnate the strata around. In many cases the bitumens occur in regular beds, which appear to have been formed in a manner similar to the deposits of true coal.

Certain important building stones have been found to be more or less impregnated with bitumen.

Such is the limestone obtained at the artesian well

quarry in the city of Chicago, and the celebrated Buena Vista, (Ohio,) sandstone used extensively in Cincinnati; also employed at Chicago in various prominent public buildings, as the post-office and Chamber of Commerce. The author, in making a chemical examination of the latter stone for the United States Treasury Department, found it to contain 2.3 per cent. bituminous matter.

OZOKERITE.

This substance, sometimes called "mineral wax," although probably a mixture of various hydrocarbons, and possibly of those belonging to two or more series, may be briefly referred to here.

In Moldavia and Galicia it is found very extensively as a brownish-yellow solid. It yields on distillation a paraffine, which is the basis of an enormous candle industry in Europe, also a wax-like body largely used in Russia as a substitute for beeswax. Very recently large deposits of black "mineral wax" have also been found in Utah, which on chemical examination the author finds to be substantially the same as the foreign, though much purer.

FOURTH SERIES.

General formula $C_n H_{2n-6}$.

BENZOL.

Synonyms ; Benzene, Benzine.
Formula $C_6 H_6$.
Sp. Gr. 0.88. Molecular weight, 78.
Sp. Gr. of vapor 2.70.
Density " " 39.
Solid at 4°. Boils at 80.5°.

Benzol is obtained, with acetylene and ethylene, in the decomposition of organic substances by heat, and its production is especially favored when the temperature is kept at a high point for some time.

Ethylene and methane form at a tolerably low temperature. Acetylene, which is richer in carbon, is produced at a higher temperature. Benzol and especially napthalin, being still more carbonaceous, are formed at an extremely high temperature.

Berthelot has prepared benzol synthetically by conducting methane tribromide, $CHBr_3$, over red-hot copper:

$$6(CHBr_3) + 9Cu = C_6H_6 + 9CuBr_2.$$

Benzol may be considered as condensed acetylene: $C_6H_6 = (C_2H_2)_3$.

Originally, benzol was prepared by a process analogous to that which furnishes methane, *i. e.*, by distilling benzoic acid with lime,

$$C_7H_6O_2 + CaO = CaCO_3 + C_6H_6.$$

At present it is obtained in immense quantities from the tar which is formed as an accessory product in the manufacture of illuminating gas.

At the high temperature of the gas-retort other products, homologous with benzol, are formed as well; viz.:

Toluene	C_7H_8	boils at	110°
Xylene	C_8H_{10}	" "	139°
Cumene	C_9H_{12}	" "	165°
Cymene	$C_{10}H_{14}$	" "	180°

and other hydrocarbides, as napthalin $C_{10}H_8$, anthracene, also various sulphur compounds, notably carbon bisulphide; several oxygenated compounds, as phenol C_6H_6O, cresylol C_7H_8O; nitrogenous compounds, as aniline C_6H_7N, and various members of its homologous series.

Benzol is a colorless, neutral liquid, with a specific gravity of 0.89, almost insoluble in water but soluble in alcohol and ether.

It dissolves sulphur, phosphorus, iodine, the different resins, and fatty substances; this latter property causes it to be employed similarly with commercial "benzine" for cleansing purposes. Care must be taken to rub with a piece of cloth having an open texture, that it may remove the benzol by absorption, without which the spot would reappear after evaporation of the solvent.

Benzol burns with a fuliginous flame. Nascent oxygen gives with it various products, and notably oxalic acid and carbon dioxide.

Chlorine and bromine yield crystalline compounds with benzol. Benzol is the simplest member of a group of bodies known as the *aromatic compounds*, of which we shall proceed to describe some of the more important.

For distinguishing benzol from the benzine of commerce, which is made from petroleum, Brandberg recommends to place a small piece of pitch in a test tube, and pour over it some of the substance to be examined. Benzol will immediately dissolve the pitch to a tar-like mass, while benzine will scarcely be colored.

NITRO-BENZOL $C_6H_5NO_2$.

This body is obtained by treating benzol with fuming nitric acid.

$$C_6H_6 + HNO_3 = C_6H_5(NO_2) + H_2O.$$

Nitro-benzol is a yellowish oil, crystallizing at 37°, has a sweet taste and an odor which has led to its use in perfumery under the name of *essence of mirbane*. Taken internally it acts as a poison.

On treatment of nitro-benzol with nascent hydrogen, hydrogen sulphide, or other reducing agent, we obtain *aniline*, which is a colorless liquid, boiling at 182°. It does not act upon litmus, yet combines with the acids, forming crystallizable compounds.

Aniline gives with chlorine, bromine and nitric acid products of substitution which are very numerous and well defined. It reacts upon the iodides of methyl, ethyl, etc., forming the corresponding *amines, or bodies constructed on the type of ammonia, having one or more of the hydrogen atoms replaced by an organic compound radicle:*

$$\text{Aniline} \quad C_6H_7N = N \begin{cases} C_6H_5 \\ H \\ H \end{cases}$$

$$\text{Methylaniline} \quad C_7H_9N = N \begin{cases} C_6H_5 \\ C\,H_3 \\ H \end{cases}$$

$$\text{Ethylmethylaniline} \quad C_9H_{13}N = N \begin{cases} C_6H_5 \\ C\,H_3 \\ C_2H_5 \end{cases}$$

C_6H_5 or, when free, $(C_6H_5)_2$, is the radicle *phenyl*, hence aniline is properly *phenylamine*.

Aniline has, during the last score of years, acquired great importance, as, under the influence of oxydizing bodies, it forms most remarkable tinctorial compounds.

If a small quantity of aniline is added to a solution of chloride of lime, the liquid is colored violet, which color disappears in a few moments. In 1858, Perkins obtained, by the action of potassium bichromate and sulphuric acid, a beautiful purple, which is known in commerce as *mauve*. Shortly after, Verguin obtained a magnificent red coloring matter on heating aniline with tin dichloride.

This substance, known under the names of *aniline-red*, *fuchsin*, *magenta*, etc., is now very economically obtained with arsenic oxide in place of the tin dichloride, which is reduced to arsenous oxide by the reaction.

Hoffmann has shown that aniline-red is a salt of a colorless base, which he calls rosaniline; this substance has the formula $C_{20}H_{21}N_3O$, or $C_{20}H_{19}N_3, H_2O$.

In the past few years there have been produced green, yellow and black colors, all originating from aniline. These substances dissolve in alcohol, and dye wool and silk without in any way weakening the fabric. They have a magnificent lustre, but their permanency is not of the highest grade.

The consumption of aniline for dyeing has now come to something enormous, amounting in Germany alone to over 15,000 tons per annum.

The aniline colors are employed in injecting tissues for microscopic preparations.

For a fuller account of the aniline colors, a larger work should be consulted.

The history of aniline affords one of the most remarkable instances of the value of scientific chemical research, when perseveringly and skillfully applied, for at first few substances seemed to promise less; and the gigantic manufacturing industry at present connected with this compound, in its applications as a tinctorial agent, offers a singular contrast to the early experiments upon this body, when a few ounces furnished a supply which exceeded the most sanguine expectations of the early discoverers of this body.

Phenol, C_6H_6O.

Synonyms: Hydrate-of phenyl, carbolic acid or phenic acid.

It occurs in castoreum, though usually procured from the portions of coal-tar distilling over between 170° and 195°. They are agitated with caustic soda, water added to separate the insoluble oils, and the phenol dissolved in the alkali is liberated as a crystalline mass, on decomposing the potassium compound with hydrochloric acid.

Salicylic acid, distilled with an excess of lime, also furnishes phenol;

$$C_7H_6O_3 + CaO = CaCO_3 + C_6H_6O.$$

If phenyl-sulphuric acid, $\left. \begin{array}{c} C_6H_5 \\ H \end{array} \right\} SO_4$, obtained by direct action of sulphuric acid upon phenol, is heated with potassium hydrate to about 300°, potassic phenol C_6H_5KO is obtained. Phenol is therefore obtained from benzol under the same conditions as alcohol is obtained from ethylene, the corresponding hydrocarbide.

Phenol crystallizes in handsome needles, fusible at 34° and boiling at 188°. It is little soluble in water, very soluble in alcohol and ether. Phenol furnishes with chlorine, bromine and iodine numerous substitution products.

Phenol has come, like alcohol, to have a generic signification, there being a number of analogous compounds, though only this, the ordinary phenol, is an important body. Heated with concentrated nitric acid, it furnishes yellow, very bitter, crystals of the body known as

Picric or Carbazotic Acid.

Picric acid is also formed when silk, benzoin, aloes, indigo, etc., are treated with nitric acid.

This acid is very largely used in dyeing, either directly to produce a yellow color, or, combined with indigo, to produce a green.

Phenol, though called carbolic acid, does not decompose the carbonates, or combine with the metals to form true salts. Phenol dissolves in sulphuric acid without coloration, if pure, and forms phenyl-sulphuric acid or sulpho-carbolic acid

$$\left. \begin{array}{c} C_6H_5 \\ H \end{array} \right\} SO_4,$$

which gives definite salts with the metals. One of these, the phenyl-sulphate or sulpho-carbolate of sodium $NaC_6H_5SO_4$, is claimed to have valuable properties as a prophylactic against scarlet fever.

Phenol gives certain reactions of the alcohols; this

somewhat explains the origin of the name given it by Berthelot. This body is the type of a class of compounds which contains:

Cresylol obtained from creosote		C_7H_8O
Phlorylol " " "		$C_8H_{10}O$
Thymol " " essence of thyme		$C_{10}H_{14}O$

PHYSIOLOGICAL ACTION OF PHENOL.

Phenol attacks the skin, producing a white stain. It coagulates albumen and is employed with great success as an antiseptic and disinfectant. It is used externally in a diluted state to dress wounds which suppurate, also in many surgical cases.

It is sometimes used internally. Large doses of it are poisonous. Carbonate and especially saccharate of calcium are considered as antidotes for phenol. Grace Calvert has announced that olive or almond oil is a still better antidote.

FIFTH SERIES.

General Formula, $C_n H_{2n-4}$.

ESSENCE, OR OIL OF TURPENTINE.

Formula $C_{10}H_{16}$.
Density of vapor compared with air 4.7.
Molecular weight, 136.
Boils at 160°

Turpentine is extracted from several varieties of the family of *Coniferæ*, notably from the pine, fir and larch.

The products vary somewhat with the nature of the tree, but they have many common characteristics; their composition is the same, their density is nearly identical and their boiling point very nearly so. Their rotary action on the solar ray varies largely.

Isomeric carbides are found in other families of plants, in the *aurantiaceæ* family for instance, as the lemons and oranges. These contain carbides very different, as evidenced by their odors and other physical properties, also different in certain chemical relations, yet having the same composition as oil of turpentine. There are also various polymers of this carbide.

This entire series of hydrocarbons can be divided into three groups. The first contains carbides having the formula $C_{10}H_{16}$, their boiling points being below 200°, and including:

	Density.	Boiling at
Oil of turpentine,	0.86	157° to 160°.
" cloves,	0.92	140° " 145°.
" lemon,	0.85	170° " 175°.
" orange,	0.83	175° " 180°.
" juniper,	0.84	about 160°.
" bergamot,	0.85	" 183°.
" pepper,	0.86	" 167°.
" elemi,	0.85	" 180°.

The carbides of the second group have the formula $C_{20}H_{32}$, their boiling point is above 200°, they are:

Oil of copaiva,	0.91	245°.
" cubebs,	0.93	240°.

The third group contains the non-volatile carbides, such as

	Density
Caoutchouc, - - - -	0.92.
Gutta-percha, - - -	0.98.

The rotary power, constant for each, varies with the different species.

French oil of turpentine causes the plane of polarization to deviate to the left; the American variety turns it 13° to the right; oil of lemon causes a deviation of 50° to the right; in the case of essence of elemi the deviation amounts to 100°. Some of the

OIL OF TURPENTINE.

essential oils of the first group contain oxygen compounds as well as the carbohydrides.

The principal chemical differences between the members of the group are the facility with which they are oxydized and their reaction with hydrochloric acid. Essence of turpentine becomes resinous rapidly when exposed to the air and finally solidifies. Essence of lemon becomes viscid after a considerable time. Hydrochloric acid produces, with essence of turpentine, a liquid and a solid compound, having each the same composition, $C_{10}H_{16}$, HCl, which, after a few weeks, becomes a dichlorhydride, (by some denominated a dichlorhydrate), $C_{10}H_{16}, 2HCl$. Essence of lemon also gives two dichlorhydrides at once, one liquid, the other solid.

Oil of turpentine may be obtained in a pure state, on distilling the commercial article in a vacuum. Thus obtained, turpentine is colorless, limpid, very volatile, and has a characteristic odor. It is insoluble in water; very soluble in alcohol and ether. It burns with a smoky flame; on exposure to the air it oxydizes and becomes resinous. The same effect is produced more rapidly with oxide of lead and some other oxides which render the oil siccative and suitable for use in painting. J. M. Merrick (100-4-289) has noticed the circumstance, important in its technical applications, that oil of turpentine attacks metallic lead quite strongly; tin, on the other hand, not at all. Turpentine, if exposed to the air, mixed with a solution of indigo, absorbs oxygen and transfers it to the indigo, which loses its color, yielding a product of oxydation called *isatin*. Under these circumstances, the turpentine does not change, and a given quantity of the essence can absorb several hundred times its volume of oxygen, and oxydize an indefinite quantity of indigo. This oxygen is probably the active modification, or ozone. Heated to 300° in a hermetically sealed tube, it changes into two products, one, isomeric, called *iso-turpentine*, which boils at 177°, and which exerts a rotatory power of 10° to 15° to the left; the other, a polymer called *meta-terebenthene*, $C_{20}H_{32}$ boiling at 360°.

OTHER SERIES OF HYDROCARBIDES.

Cinnamene C_8H_8 is a very refractive liquid with a density of 0.924, boiling at 146°. *Styrol* which is produced from storax is converted at 205°, into a polymeric solid, termed *Meta-styrol* or *Draconyl*. If styrol is made to act upon acetylene, or ethylene, at a red heat, there is obtained the very important hydrocarbide *naphthalin* $C_{10}H_8$. This is a body crystallizable in very handsome plates, and is ordinarily obtained from coal tar by distillation between 200° and 300°; heavy oils pass over, out of which naphthalin crystallizes; on cooling, the mass is pressed and purified by sublimation. It fuses at 79° and distils at 220°.

Naphthalin is associated in coal tar with a hydrocarbide, beautifully crystallizing in long needles, fusing at 93° and boiling at 285°. This is *acenaphtene*

$C_{12}H_{10}$. Another hydrocarbide is also found in this tar, *anthracene*. Its formula is $C_{14}H_{10}$. It forms very diminutive crystalline plates fusing at 210° and boiling at 360°. Its vapor is extremely acrid.

This body has recently enabled chemists to reproduce the coloring principle of madder; *alizarin* $C_{14}H_8O_4$. It is obtained on oxydizing anthracene by means of a mixture of bichromate of potassium and sulphuric acid, which gives *oxyanthracene* $C_{14}H_8O_2$. This, with fused potassa, furnishes a combination of potassium and alizarin, from which the latter is precipitated by an acid. It has the form of brilliant bronze-colored needles, identical with natural alizarin obtained from madder.

Alizarin sublimes at 215° and is very stable, little soluble in cold water, but readily soluble in boiling water. It is easily dissolved in alcohol, ether and carbon bisulphide.

Its chemical character, not quite well defined as yet, appears to place it among the phenols. (See page 33.)

The artificial production of alizarin from anthracene, thus furnishing a cheap substitute for madder, the chief dye-stuff used in printing calicoes, is one of the latest and most noteworthy triumphs of organic chemistry. Thousands of acres of land in Europe, especially in Alsatia, now devoted to the culture of madder, may be restored to cereal or other food agriculture.

Before leaving the hydrocarbons proper, it should

be stated that compounds of carbon and hydrogen of extra-terrestrial origin have been found in certain meteorites, by J. Lawrence Smith. (80-76-388.)

CAMPHOR.

Camphor is usually considered at this point, on account of its intimate relation to the oxydized essential oils in composition, and to turpentine in many chemical reactions.

Berthelot regards camphor as an aldehyd. Kekulé places it among the ketones.

Camphor exists in various parts of the *Laurus camphora*. To obtain it, the wood is finely divided and heated with water in a metallic vessel, closed by a cover filled with straw. The camphor is condensed in grayish crystals on the straw, forming the crude camphor of commerce; it is afterwards sublimed in a glass retort as a further purification.

Camphor is a crystallized body, having a burning taste and an aromatic odor. Its density is 0.99 at 10°. It is elastic and with difficulty pulverized, which can, however, be easily effected on moistening with a few drops of alcohol. Water dissolves only about $\frac{1}{1000}$ part of it; thrown upon pure water it floats on the surface with a gyratory motion. It is soluble in alcohol, ether, acetic acid and essential oils; it is sublimed at ordinary temperatures where kept in close vessels, and deposits again on the cooler side of the receptacle.

It burns with a smoky flame and oxydizes on being

boiled with nitric acid, yielding *camphoric acid* $C_{10}H_{16}O_4$ which is bibasic. Heated with zinc chloride or anhydrous phosphoric acid, it furnishes *Cymol* $C_{10}H_{14}$.

The author found (1-146-73) that on treatment of camphor with hypochlorous acid he obtained the new body, $C_{10}H_{15}ClO$, which he denominates *monochlor-camphor*; this, on treatment with alcoholic potassium hydrate, yielded *oxycamphor* $C_{10}H_{16}O_2$.

Camphor is very extensively employed in medicine and pharmacy.

RESINS, BALSAMS, GUM-RESINS.

These bodies are products of the oxidation of essential or volatile oils. The name of *gum-resin* is applied to those which contain a gum, and *balsam* to those which contain essential oils and an acid, usually cinnamic or benzoic, in addition to the resin which is presented in both. A. B. Prescott, the eminent authority on proximate analysis, defines balsams as "natural mixtures of volatile oils with their oxidation products,—resins and solid volatile acids."

They are substances more or less colored, hard and brittle. They are fusible, non-volatile, and burn with a fuliginous flame. They are insoluble in water, generally soluble in alcohol, ether and essential oils.

Several of them are acid. This is the case with the most important of them, as the resin of the pine, called *colophony*, from which three isomeric acids have been obtained—the *pinic*, *sylvic*, and *pimaric*, $C_{20}H_{30}O_2$.

ORGANIC CHEMISTRY.

This resin constitutes the fixed residue obtained on distilling crude turpentine. It is used for preparing varnish, in soldering, and in certain combinations with the alkalies, called resin-soaps.

Subjoined are given the names and the origin of the principal resins, oleo-resins, gum-resins and balsams. With some, the position assigned them in this classification is not definitely settled.

RESINS.

Amber is found in the lignites and in the alluvial sands of the Baltic.

Arnicin, the active principle of Arnica Root.

Cannabin, the active principle of *Indian Hemp*.

Castorin, a secretion of the Beaver (*Castor*).

Ergotin(?), the active principle of Ergot of common rye.

Mastic, a resinous exudation of the Mastic, or Lentisk tree.

Burgundy Pitch, an exudation of the Spruce Fir, *Abies excelsa*.

Pyrethrin, the active principle of the Pellitory root.

Rottlerin, a crystalline resin from Kamala, the minute glands which cover the capsules of *Rottlera tinctoria*.

OLEO-RESINS.

Copaiva, a resinous juice of the *copaifera officinalis* found in Spanish America.

Wood-oil, an oleo-resin from the *Dipterocarpus turbinatus*.

RESINS, BALSAMS, GUM-RESINS. 43

Elemi, an exudation of an unknown tree, (probably *Cannarium commune*).

Common Frankincense, a concrete turpentine of the *Pinus tæda*.

Canada balsam, the turpentine of the Balm of Gilead Fir, (*Abies balsamea*).

Storax, from the *Liquidambar orientale*.

GUM-RESINS.

Ammoniacum, an exudation of the *Dorema ammoniacum*.

Assafœtida, a gum resin obtained by incision from the living root of the *Narthex assafœtida*.

Gamboge, obtained from the *Garcinia morella*.

Galbanum, from the *Ferula galbaniflua*.

Myrrh, an exudation of the *Balsamodendron myrrha*.

BALSAMS.

Benzoin, obtained from incisions of the bark of *Styrax benzoin*.

Balsam of Peru, from the *Myroxylon·Pereiræ*.

Balsam of Tolu, obtained from incisions of the bark of *Myroxylon toluifera*.

Caoutchouc is the hardened juice of *Ficus elastica*, *Jatropha elastica*, *Siphonia cahuchu*, and other plants.

Gutta-percha is the concrete juice of the *percha* (Malay) tree the *Isonandra percha*, a sapotaceous plant.

ALCOHOLS.

GENERAL DEFINITION AND CHARACTERISTICS.

This name is given to a class of neutral bodies as important as they are numerous. Their essential characteristic is that of reacting upon acids so as to form water and a class of bodies called *ethers*.

The number of alcohols is very considerable. There are several distinct varieties of alcohol recognized.

I. Those built on the type of one molecule of water:

$$\left.\begin{array}{c} C_2H_5' \\ H \end{array}\right\} O, \text{ ethyl or common alcohol.}$$

II. On two molecules of water:

$$\left.\begin{array}{c} (C_2H_4'' \\ H_2 \end{array}\right\} O_2, \text{ ethylene alcohol or glycol.}$$

III. On three molecules of water:

$$\left.\begin{array}{c} C_3H_5''' \\ H_3 \end{array}\right\} O_3, \text{ glycerine and thus on.}$$

They may be defined as bodies built on the type of one or more molecules of water having one-half of the hydrogen replaced by a hydrocarbide radicle.

MONATOMIC ALCOHOLS,

or those formed on the type of one molecule of water,

ALCOHOLS.

of which ordinary alcohol is the best studied, are characterized by the fact that they contain one atom of oxygen only, and that by reaction with the monobasic acids they form but a single ether.

They may be obtained synthetically, as well as by various indirect processes.

Subjoined is a classified list of the more important monatomic alcohols:

FIRST SERIES,

$$C_nH_{2n+2}O.$$

Methyl alcohol (wood spirit),	$C H_4 O$
Ethyl alcohol, (spirit of wine)	$C_2 H_6 O$
Propyl alcohol	$C_3 H_8 O$
Butyl alcohol,	$C_4 H_{10} O$
Amyl alcohol,	$C_5 H_{12} O$
Setyl alcohol	$C_6 H_{14} O$
Octyl alcohol	$C_8 H_{18} O$
Sexdecyl alcohol	$C_{16} H_{34} O$
Ceryl alcohol	$C_{27} H_{56} O$
Myricyl alcohol	$C_{30} H_{62} O.$

SECOND SERIES,

$$C_nH_{2n}O.$$

Vinyl alcohol	$C_2 H_4 O$
Allyl alcohol	$C_3 H_6 O.$

THIRD SERIES,

$$C_n H_{2n-2} O.$$

Borneol alcohol	$C_{10}H_{18}O.$

FOURTH SERIES,

$$C_nH_{2n-6}O.$$

Benzyl alcohol	C_7H_8O
Xylyl alcohol	$C_8H_{10}O$
Cumol alcohol	$C_9H_{12}O$
Cymol alcohol	$C_{10}H_{14}O$

FIFTH SERIES,

$$C_nH_{2n-8}O.$$

Cinnyl alcohol	$C_9H_{10}O$
Cholesteryl alcohol	$C_{26}H_{44}O$

MONATOMIC ALCOHOLS HAVING THE GENERAL FORMULA,

$$C_nH_{2n+2}O.$$

METHYL ALCOHOL, OR WOOD-SPIRIT.

$$CH_4O = \left.\begin{matrix} CH_3 \\ H \end{matrix}\right\} O.$$

This substance is found in the liquid obtained on distilling wood. The distillate contains in addition, water, acetic acid, tar, and various oils. In order to extract the methyl alcohol, it is again distilled and that portion which passes over at 90° is collected; this is diluted with water, the oil which precipitates separated, and the liquid agitated for a considerable time with olive oil. This oil is then removed, the liquid redistilled several times and only that portion collected which passes over above 70°. On being again distilled with calcium chloride it furnishes methyl alcohol, nearly pure, boiling at 66.5°.

There are other methods of rectifying besides the one here given.

This body possesses most of the general properties of ordinary alcohol. Under the action of the oxides it furnishes an aldehyd and formic acid.

With the acids it produces ethers; viz., with hydrochloric acid, methyl chloride, $CH_3Cl = \left. \begin{matrix} CH_3 \\ Cl \end{matrix} \right\}$;

with acetic acid,

methyl acetic ether, $C_3H_6O_2 = \left. \begin{matrix} CH_3 \\ C_2H_3O \end{matrix} \right\} O$.

CHLOROFORM, $CHCl_3$.

Methyl chloride produces with chlorine a regular series of products of substitution. One of these terms, $CHCl_3$, is the very important body, *chloroform*, discovered in 1831 by Soubeiran and Liebig.

To prepare this compound, 40 litres of water, 5 kilos of recently slacked lime, and 10 kilos of chloride of lime are heated to 40°; 1500 grams of 90 per cent. alcohol are then added and the retort luted with clay.

It is now heated for a moment to the boiling point and the fire then at once slackened.

The ebullition having ceased there will be found two layers in the receiver. The upper layer is formed of water and alcohol, the lower one is chloroform nearly pure. The latter is washed with water, agitated with a dilute solution of potassium carbonate, or with fused calcium chloride for twenty-four hours, and distilled to four-fifths.

Chloroform is a colorless liquid. When first prepared it has a sweetish penetrating taste, and an agreeable, ethereal odor.

Its density is 1.48; it boils at 60.5°, is soluble in alcohol and ether and difficultly so in water.

It burns, though not readily; its flame having a green margin. It dissolves iodine, sulphur, phosphorus, fatty substances and resins.

An alcoholic solution of potassa decomposes it into chloride and formiate:

$$CHCl_3 + 4KHO = 3KCl + CHKO_2 + 2H_2O.$$

Physiological Action.

Chloroform is at present very generally used as an anesthetic. Opinions as to its manner of acting are divided. Formerly it was thought that the insensibility produced was the commencement of asphyxia. Since then it has been ascertained that the heart, in case of poisoning by chloroform, immediately loses all power of contraction, and it is now generally admitted that paralysis of the muscles and nerves of the heart is produced.

As the vapor of chloroform is very dense, care should be taken that in its use, access of air to the lungs be not wholly prevented, or serious consequences may result. Probably the fatal accidents that have occurred may, in some instances at least, be attributed to lack of care in this regard.

It is of great importance that the chloroform used should be quite pure. In some cases it has been found to have undergone spontaneous decomposition after exposure to a strong light. It ought to communicate no color to oil of vitriol when agitated with it. The liquid itself should be free from color or any chlorous odor. When a few drops are allowed to evaporate on the hand no unpleasant odor should remain.

Shuttleworth (100, 4, 339) states that partially decomposed chloroform can be rectified by agitating it with a solution of sodium hypo-sulphite.

ORDINARY ALCOHOL.

Ethylic, or Vinic Alcohol.

Formula: C_2H_6O.
Density of vapor 23.
Density .81.
Boils at 78.4°.
Cannot be solidified.

It is prepared by the fermentation of saccharine liquids at a temperature of 25° to 30°, in the presence of a small quantity of a ferment. Cane sugar does not *directly* become alcohol under the influence of a ferment. It is first transformed into two other sugars, glucose and levulose.

$$C_{12}H_{22}O_{11} + H_2O = \underbrace{C_6H_{12}O_6}_{\text{Glucose.}} + \underbrace{C_6H_{12}O_6}_{\text{Levulose.}}$$

In its final fermentation nearly all the sugar is changed into alcohol and carbon dioxide,

$$C_6H_2O_6 = 2C_2H_6O + 4CO_2.$$

This equation accounts for the transformation of 94 to 96 per cent. of the sugar employed, but besides alcohol and carbon dioxide, succinic acid is always formed as well as glycerine, and in most cases " fusel oil," consisting chiefly of amyl alcohol.

Fermentation is a phenomenon correlative with the development and growth of cells of the fungus *Mycoderma* (*Torula*) *cerevisiæ* which constitutes yeast. Sometimes the sugar is furnished as a natural product by fruits; often glucose is produced from the starch of cereals, potatoes, etc., and then changed into alcohol afterwards. Corn is the leading original source in this country.

Alcohol obtained by fermentation is concentrated by distillation. This operation is performed in retorts, the construction of which is based upon a principle developed by A. de Montpellier, and improved by Derosne, Dubrunfaut and others. The object is to prevent the distilling over of the water with the alcohol, and is quite well accomplished by the improved methods now employed. The details are not suited to the scope of this work.

The application of this rational method of distilling

admits of obtaining liquids containing up to 90 per cent. of alcohol, but it is difficult to go beyond that point of concentration.

In order to prepare alcohol more concentrated, substances having a great avidity for water must be used. Calcium chloride is not suitable, as it unites with the alcohol. Anhydrous sulphate of copper, carbonate of potassium or quicklime do not produce absolute alcohol. But it is very rare that perfectly anhydrous alcohol is required. Alcohol of 97 per cent. is obtained in treating alcohol of 85 per cent. during two days with lime, or better, with a sixth or seventh part of its weight of dry potassium carbonate, and then distilling. If it is desired to procure absolute alcohol, very concentrated alcohol is treated with caustic baryta until the liquid is colored yellow and then distilled.

Alcohol in fresh bread made with yeast has been found by Bolas (8-27-271) to the amount of .314 per cent. Slices of bread a week old contained .12 to .13 per cent.

Absolute alcohol is a colorless liquid, more limpid than water, of an agreeable odor and a burning taste. It boils at 78.4°, is neutral, combustible and burns with a flame but little luminous. It heats on coming in contact with water, and attracts the moisture of the air very rapidly.

It contracts upon mixing with water; the maximum of contraction takes place at a temperature of 15° when 52.3 vol. of absolute alcohol are mixed with 47.7 vol. of water; instead 100 vol. one obtains

96.3 vol. At the moment of admixture numerous air bubbles escape and the mixture becomes heated.

The alcoholic strength of the liquids consumed as beverages varies considerably.

Madeira wines, about			20 per cent.	
Malaga	"	"	14 to 16	"
Bordeaux	"	"	15 to 12	"
Rhine	"	"	10 to 12	"
California	"	"	10 to 16	"
Cider	"	"	2 to 7	"
Beer	"	"	1 to 8	"

Spirits are distilled from fermented liquids; *brandy* from wine; *whisky* from a mash of corn or rye; *rum* from molasses, etc. They contain about 50 per cent. of alcohol.

The term *proof spirits* was originally given to alcohol sufficiently strong to fire gunpowder when lighted. The strength of proof spirits now varies in different localities, and it would be well were this ambiguous designation no longer employed.

Alcohol dissolves the caustic alkalies, certain nitrates, chlorides and other salts, also various gases. With some of these, genuine chemical combinations are produced, and not mere solutions; this is the case with calcium chloride and magnesium nitrate. Alcohol can be mixed with ether in all proportions; it dissolves the resins, essential oils, and a great number of other organic bodies.

The chemical properties of alcohol are very inter-

ALCOHOLS. 53

esting. Vapor of alcohol is decomposed on passing through a tube heated to redness; hydrogen, marsh-gas, oxide of carbon, small quantities of naphthalin, benzol, and phenol are formed. In presence of air and water it slowly oxidizes and yields acid compounds. This action is rapid, if a hot spiral of platinum is placed in the alcoholic vapor.

EXPERIMENT.—Place a small platinum spiral in the wick of an alcohol lamp, light and then blow out the flame. It will be seen that the spiral remains incandescent. Spongy platinum acts still more energetically; if very concentrated alcohol is poured drop by drop into a capsule containing spongy platinum, or platinum black, it will be seen to redden, fumes are produced and an acid liquid is formed containing chiefly aldehyd and acetic acid. The same oxidation occurs if diluted alcohol is exposed to the air in the presence of *mother of vinegar*, a cryptogamic plant. (*Mycoderma aceti*). In fact, this is the basis of the manufacture of wine-vinegar and alcohol.

Fuming nitric acid reacts upon alcohol with explosive energy. Aldehyd is formed, also acetic ether, nitrous ether and acetic, formic, glycollic, oxalic and carbonic acids. Alkaline hydrates attack alcohol even in the cold potassium acetate being the chief product formed. If alcoholic vapor is made to pass over lime heated to 250°, hydrogen gas and calcium acetate are produced; the latter is decomposed at a more elevated temperature into marsh gas and water. If silver or mercury is dissolved in nitric acid, and 90 per cent. alcohol added to the cooled solutions, a lively ebullition results, and a crystalline precipitate is deposited which explodes at 185°, or by percussion. This body is the *fulminate of silver or mercury*, respectively, which is considered as derived from methyl cyanide, CH_3Cy, by the substitution of 1 molecule of nitryl, and of 1 atom of mercury, or 2 of silver for 3 atoms of hydrogen. The formulæ are $C(NO_2)HgCy$; $C(NO_2)Ag_2Cy$.

Potassium attacks absolute alcohol, and is dissolved liberating hydrogen; on cooling, potassium ethylate is deposited. Sodium acts in the same manner. These compounds, if brought in contact with water, regenerate alcohol and the respective alkaline hydrates.

Acids attack alcohol and furnish compound ethers, which we will study later. Ozone, according to A. W. Wright, (80—[3]7—184) oxydizes alcohol to acetic acid.

PHYSIOLOGICAL ACTION OF ALCOHOL. USES OF ALCOHOL.—Alcohol coagulates the blood; injected into the veins it produces instantaneous death. It is a very powerful poison, as are all alcohols of the series $C_n H_{2n+2} O$. Rabuteau (9—81—631) has shown that they are more poisonous in proportion as their molecules are complex. Cases have been observed where a large dose of alcohol has caused death in half an hour.

The worse than worthless character of distilled liquors as beverages is no longer an open question. With regard to their value as food or medicine, a more authoritative or competent expression of opinion cannot be desired than that of the International Medical Congress, which at its session in Philadelphia in 1876, said:

ALCOHOLS.

"1. Alcohol is not shown to have a definite food value by any of the usual methods of chemical analysis or physiological investigation.

"2. Its use as a medicine is chiefly that of a cardiac stimulant, and often admits of substitution.

"3. As a medicine, it is not well fitted for self-prescription by the laity, and the medical profession is not accountable for such administration, or for the enormous evils arising therefrom.

"4. The purity of alcoholic liquors is, in general, not as well assured as that of articles used for medicine should be. The various mixtures when used as medicine, should have definite and known composition, and should not be interchanged promiscuously."

The dissolving power of alcohol renders it very serviceable in the arts. Solutions in this menstruum are called *alcoholic tinctures*. Only the purest alcohol ought to be used in pharmacy, though of course, various strengths are requisite, as it should be of a degree to suit the nature of the matter to be dissolved. If the substance to be treated is a resin, or some substance absolutely insoluble in water, a very concentrated alcohol is preferable. A weaker alcohol is made use of, if the matter is one that is soluble, both in alcohol and water.

Alcohol acts not only as a solvent, but also as a preventative of decay. This is a property which renders it especially valuable in the preparation of remedies.

AMYL ALCOHOL.

$$C_5H_{12}O = \left. \begin{array}{c} C_5H_{11} \\ H \end{array} \right\} O.$$

Synonyms: Fousel (or Fusel) Oil, Potato Spirit.

The amylic compounds derive their name from *Amylum*, starch, the chief constituent of the potato. They are formed in some proportion in almost every instance of alcoholic fermentation of sugar. Amylic alcohol is usually prepared on fractionally redistilling the oil which remains when the alcohol, prepared from potatoes, barley, corn, etc., is distilled. The product which comes over at 132°, is that collected. Cahours and Balard first established the analogy, in constitution and properties, of this compound with ordinary alcohol. It is a monatomic alcohol, giving with oxidizing re-agents, *valeric acid*.

$$\underbrace{C_5H_{12}O+O_2}_{\text{Amylic alcohol.}} = \underbrace{C_5H_{10}O_2+H_2O.}_{\text{Valeric acid.}}$$

and with acids, compound ethers, as

Chloride of amyl, $\quad C_5H_{11}Cl.$

Acetate of amyl or amyl-acetic ether, $\left. \begin{array}{c} C_5H_{11} \\ C_2H_3O \end{array} \right\} O.$

MONATOMIC ALCOHOLS.

Having the general Formula $C_nH_{2n}O$.

ALLYLIC ALCOHOL, $C_3H_6O = \left.\begin{array}{c} C_3H_5 \\ H \end{array}\right\} O.$

This is a body giving the same reactions as ordinary alcohol. The radicle it contains is the same as that in the triatomic alcohol, glycerine. Among its derivatives there are two which are of considerable importance:

Allyl sulphide, $\left.\begin{array}{c} C_3H_5 \\ C_3H_5 \end{array}\right\} S.$

Sulpho-cyanide, $\left.\begin{array}{c} C_3H_5 \\ C\ N \end{array}\right\} S.$

The former is oil of garlic; the latter oil of mustard. OIL OF GARLIC is prepared by the following method: allylic alcohol is treated with phosphorus iodide which furnishes allyl iodide C_3H_5I. This iodide is afterwards mixed with an alcoholic solution of potassium sulphide and the whole is distilled; the product which passes over is identical with the essential oil obtained in distilling garlic, onions, assafœtida, etc., with water.

OIL OF MUSTARD, OR SULPHO-CYANIDE OF ALLYL.

This body is prepared by causing iodide of allyl to react upon potassium sulpho-cyanide, $\left.\begin{array}{c} CN \\ K \end{array}\right\} S.$ and may be regarded as sulpho-cyanic acid, $\left.\begin{array}{c} CN \\ H \end{array}\right\} S,$ having the hydrogen replaced by the radicle of allyl alcohol, C_3H_5. The product which distills over is an irritating liquid which boils at 145°, like the oil prepared from mustard directly. This substance may also be obtained by the action of allylic alcohol upon potassium sulpho-cyanide. It is likewise obtained by the fermentation of mustard seeds.

Sulpho-cyanide of allyl does not exist already formed in black mustard (*Sinapis nigra*), but according to Bussy, its formation is due to a particular ferment.

Oil of mustard combines directly with ammonia, forming a crystalline substance called *thiosinnamine*. $C_4H_8N_2S$, which, in contact with mercuric oxide, changes into an alkaloid called *sinnamine*, of which the composition is $C_4H_6N_2$. It reacts upon lead oxide producing a substance called *sinapoline* whose formula is $C_7H_{12}N_2O$.

BORNEO CAMPHOR, OR BORNEOL $C_{10}H_{18}O$.

This body exudes from the *Dryobalanops camphora* (Borneo). It is crystalline and has an odor between that of camphor and pepper. It fuses at 195°, and boils at about 220°. It is dextrogyrate. Heated with nitric acid it furnishes common camphor $C_{10}H_{16}O$.

DIATOMIC ALCOHOLS OR GLYCOLS.

$$C_nH_{2n+2}O_2.$$

Ordinary Glycol, $(C_2H_4) - O_2 - H_2 = C_2H_6O_2$
Propyl " $(C_3H_6) - O_2 - H_2 = C_3H_8O_2$

ALCOHOLS.

Butyl Glycol, $(C_4H_8) - O_2 - H_2 = C_4H_{10}O_2$
Amyl " $(C_5H_{10}) - O_2 - H_2 = C_5H_{12}O_2$
Hexyl " $(C_6H_{12}) - O_2 - H_2 = C_6H_{14}O_2$
Octyl " $(C_8H_{16}) - O_2 - H_2 = C_8H_{18}O_2$.

TRIATOMIC ALCOHOLS.

Glycerine, $(C_3H_5) - O_3 - H_3 = C_3H_8O_3$.

TETRATOMIC ALCOHOLS.

Erythrite, $(C_4H_6) - O_4 - H_4 = C_4H_{10}O_4$.

OTHER COMPLEX ALCOHOLS.

Glucose and its isomerides, $(C_6H_6) - O_6 - H_6 = C_6H_{12}O_6$.
Mannite, - - $(C_6H_8) - O_6 - H_6 = C_6H_{14}O_6$.
Dulcite, - - - $(C_9H_8) - O_6 - H_6 = C_6H_{14}O_6$.
Quercite,
Pinnite, $\left.\right\} C_3H_6O + \left.\begin{matrix}(CH_2)_2\\H_2\end{matrix}\right\} O_2 + C_6H_{12}O_5$.

ORDINARY GLYCOL.

$$C_2H_6O_2 = \left.\begin{matrix}(CH_2)_2\\H_2\end{matrix}\right\} O_2.$$

The discovery of the glycols was an event of great importance. It was achieved by Wurtz in 1856, and the glycol of which we are treating was the first discovered.

In a flask surmounted by a condenser, two parts of potassium or sodium acetate, are dissolved in weak alcohol and one part of ethylene bromide added. This mixture is heated in a water bath as long as the precipitate of alkaline bromide continues to form, care being taken at the same time to keep the worm well cooled, in order that the vapors of alcohol may continually flow back into the flask. The alcohol is distilled off in a water bath, and the residue afterwards also distilled at a higher temperature, and that part collected which passes over between 140° and 200°. This portion which contains monacetic glycol, is heated with a saturated solution of baryta until the liquid acquires a strong alkaline reaction. The excess of baryta is removed by passing carbon dioxide through the solution which is then filtered and evaporated. The barium acetate is precipitated completely by strong alcohol, and the alcohol subsequently removed by distillation. The retort is now heated in an oil bath, and that portion set aside which boils above 150°. This is redistilled and the distillate between 190° and 198° is the product sought. Zeller and Huefner have lately (18, 10,270) obtained the purest glycol by simply heating a solution of potassium carbonate with ethylene bromide.

Glycol is a colorless, odorless liquid. somewhat viscid and having a sweetish taste. Its density is 1.12; water and alcohol dissolve it in all proportions. Ether dissolves it with difficulty.

It is not oxydized in the air under ordinary conditions, but if dilute glycol be made to fall on platinum black, it becomes heated and is transformed into *glycolic acid*. Its equivalence is shown by the follow-

ALCOHOLS.

ing: glycol attacks sodium forming two sodium glycols;

$$\left.\begin{array}{l}C_2H_4\\NaH\end{array}\right\}O_2, \quad \left.\begin{array}{l}C_2H_4\\Na_2\end{array}\right\}O_2.$$

These glycols furnish two ethyl glycols on being heated with ethyl iodide.

$$\left.\begin{array}{l}C_2H_4\\C_2H_5,H\end{array}\right\}O_2, \quad \left.\begin{array}{l}C_2H_4\\(C_2H_5)_2\end{array}\right\}O_2.$$
Ethyl-glycol.   Diethyl-glycol.

With hydrogen bromide it furnishes two different products according to the number of molecules of HBr taken.

$$C_2H_6O_2 + HBr = \underbrace{C_2H_5BrO}_{\text{Monobromhydric ether.}} + H_2O.$$

$$C_2H_6O_2 + 2HBr = \underbrace{C_2H_4Br_2}_{\text{Ethylene bromide.}} + 2H_2O.$$

It is evident that mixed ethers may be obtained by treating glycol not with two molecules of the same acid, but with two molecules of different acids. Thus aceto-chlorhydric glycol is formed $\left.\begin{array}{l}C_2H_4\\(C_2H_3O)Cl\end{array}\right\}O.$

These ethers, in the presence of alkalies, are reformed into their respective acids and glycol, in the same manner in which ethers of ordinary alcohol regenerate alcohol.

Monochlorhydric and aceto-chlorhydric glycol form an exception to this rule; they form oxide of ethylene in presence of alkalies.

OXIDE OF ETHYLENE, C_2H_4O,

a polymer of $(C_2H_4)_2O_2$, is related to glycol as ordinary ether to alcohol. It is not obtained like the latter by the action of hydrogen sulphate on the alcoholic compound, but is produced by the action of potassa on monochlorhydric glycol. A solution of potassa is gradually poured into chlorhydric glycol placed in a glass, or a tubulated retort.

$$KHO + C_2H_5ClO = KCl + H_2O + C_2H_4O.$$

The oxide of ethylene distills over with the water; the latter is absorbed by causing the vapors to pass through a flask containing anhydrous calcium chloride, and the oxide is condensed in a receptacle placed in a refrigerating mixture.

It is a colorless, ethereal, fragrant liquid; boiling at 13°. Its density is 0.89. Ethylene oxide is very soluble in water, alcohol and ether. It burns with a luminous flame and reduces silver salts. It has the composition but not the properties of aldehyd, of which it is an isomeride.

Oxide of ethylene is a very remarkable body. It combines directly with oxygen, hydrogen, chlorine and bromine, also combines directly with acids, often even with the disengagement of heat, forming the ethers of glycol and polyethylenic alcohols. This body is therefore a true non-nitrogenous basic oxide.

TRIATOMIC ALCOHOLS OR GLYCERINES.

Ordinary Glycerine, $C_3H_8O_3 = \begin{matrix} C_3H_5 \\ H_3 \end{matrix} \Big\} O..$

This body, discovered by Scheele, in 1779, and called by him, on account of its sweet taste, *the sweet principle of oils,* has been specially studied by Chevreul and by Pelouze. Berthelot discovered its real nature and proved it to be a triatomic alcohol.

Glycerine is prepared by decomposing neutral fatty bodies, in the soap and candle industry by alkalies, or better still by superheated steam. (*Tilghman's process.*) It is obtained in pharmacy, whenever lead plaster is prepared and remains in the water with which the latter is washed.

It is much employed in pharmacy and perfumery and as a solvent for many substances. Crude glycerine may be purified by boiling with animal charcoal and filtering before being evaporated to the required consistency. The best process consists in distilling the crude condensed glycerine in a current of steam. Pasteur has shown that glycerine is produced in a very small quantity in alcoholic fermentation. We owe to Wurtz, a remarkable synthetical reproduction of glycerine. Propylene C_3H_6 furnishes an iodide C_3H_5I, called iodide of allyl. This body produces with bromine the

ALCOHOLS.

compound $C_3H_5Br_3$ which, treated with potassa, or oxide of silver, yields glycerine.

$$C_3H_5Br_3 + 3KHO = 3\ KBr. + \underset{\text{Glycerine.}}{C_3H_8O_3}.$$

Glycerine is a syrupy liquid, colorless, of a sweetish taste and destitute of odor; its density is 1.28 at 15°. Sarg has obtained crystals of glycerine, whose angles have been measured by Victor Lang (2-152-637). They are rhombic in form and very deliquescent. Glycerine is soluble in alcohol and water in all proportions; it is not dissolved by ether. It dissolves alkalies, alkaline sulphates, chlorides and nitrates, copper sulphate, silver nitrate and many other salts.

Glycerine distills at 280°, but is thereby partially decomposed. It may, however, be distilled in a vacuum without change. It is decomposed at a temperature above 300°, and oils, inflammable gases, carbon dioxide, and a product very irritating to the eyes, called *acrolein*, acrylic aldehyd, are formed: this last substance may be obtained pure by distilling glycerine with sulphuric, or phosphoric acid. The formula of acrolein is $C_3H_4O_2$; it is also produced in the dry distillation of all fatty bodies which contain glycerine. If glycerine be made to fall drop by drop upon platinum black, it unites, like alcohol and glycol, with O_2 and *glyceric acid* is formed.

$$C_3H_8O_3 + O_2 = C_3H_6O_4 + H_2O.$$

The oxidation of the glycerine does not stop here;

there is subsequently formed, acetic, formic, and carbonic, but chiefly oxalic acid. The action of acids on glycerine demonstrates two facts; first, that glycerine is an alcohol; second, that it is a triatomic alcohol. On treating glycerine with hydrochloric acid the first reaction is similar to that between alcohol and this acid,

$$HCl + C_3H_8O_3 = C_3H_7ClO_2 + H_2O.$$
Monochlorhydric ether,
or
Monochlorhydrin.

The continued action of phosphorous perchloride upon glycerine, or the dichlorhydrate of glycerine, effects the elimination of additional molecules of water and the formation of trichlorhydrin.

$$3HCl + C_3H_8O_3 = C_3H_5Cl_3 + 3(H_2O).$$
Trichlorhydrin.

Berthelot has studied the acetines, butyrines (tributyrine exists in butter), valerines, and many other ethers of glycerine. If glycerine is mixed with cold nitric acid, and sulphuric acid added drop by drop, an oily substance separates out which is *trinitroglycerine*, $C_3H_5(NO_2)_3O_3$. This body detonates with great violence. It acts very energetically on the system. A few drops placed on the tongue produce violent megrim. Glycerine forms compounds with lime analogous to those formed by sugar, according to P. Carles, (1-174-87).

USES.—The uses of glycerine in the arts, and especially in pharmacy, are numerous and important, many of which are based upon the solvent power of this compound. Henry Wurtz (31-195-58) has made valuable suggestions as to its economical applications.

TABLE SHOWING THE SOLUBILITY OF SOME CHEMICALS IN GLYCERINE, (FROM KLEVER.) ONE HUNDRED PARTS OF GLYCERINE DISSOLVE THE ANNEXED QUANTITIES OF THE FOLLOWING CHEMICALS:

Arsenous oxide,	20.00
Arsenic oxide,	20.00
Acid, benzoic,	10.00
" oxalic,	15.00
" tannic,	50.00
Alum,	40.00
Ammonium carbonate,	20.00
" chloride,	20.00
Antimony and potassium tartrate,	5.50
Atropia,	3.00
Atropia sulphate,	33.00
Barium chloride,	10.00
Brucia,	2.25
Cinchonia,	0.50
" sulphate,	6.70
Copper acetate,	10.00
" sulphate,	30.00
Iron and potassium tartrate,	8.00
" lactate,	16.00
" sulphate,	25.00
Mercuric chloride,	7.50
Mercurous chloride,	27.00
Iodine,	1.90
Morphia,	0.45
Morphia acetate,	20.00
" chlorhydrate,	20.00
Phosphorus,	0.20
Plumbic acetate,	20.00
Potassium arsenate,	50.00
" chlorate,	3.50
" bromide,	25.00
" cyanide,	32.00
" iodide.	40.00
Quinia,	0.50
" tannate,	0.25

Sodium arsenate.	50.00
" bicarbonate,	8.00
" borate,	60.00
" carbonate,	98.00
" chlorate,	20.00
Sulphur,	0.10
Strychnia,	0.25
" nitrate,	4.00
" sulphate,	22.50
Urea,	50.00
Veratria,	1.00
Zinc chloride,	50.00
" iodide,	40.00
" sulphate,	25.00

The general use of glycerine in pharmacy, to prevent solid extracts from becoming too hard by evaporation, is greatly to be deprecated, as an adulterant is thereby introduced, which renders this class of remedies more or less unreliable.

ETHERS.

SIMPLE ETHERS.

Ethers are products formed by the action of alcohols upon acids.

By most chemists they are looked upon as referable to the oxides of metals; thus $\begin{matrix}CH_3\\CH_3\end{matrix}\Big\}O$ and $\begin{matrix}C_2H_5\\C_2H_5\end{matrix}\Big\}O$, may be regarded as the oxides respectively of methyl and ethyl. They bear the same relation to alcohols that oxides of the metals do to the hydrates.

Potassium hydrate	KOH.
Ethyl hydrate, or ethyl alcohol	C_2H_5OH.
Potassium oxide	$\begin{matrix}K\\K\end{matrix}\Big\}O.$
Ethyl oxide or ethyl ether	$\begin{matrix}C_2H_5\\C_2H_5\end{matrix}\Big\}O.$

The simple ethers are mostly liquid. They are very slightly soluble in water, while they are readily soluble in alcohol. Exposed to the action of alkaline solutions they regenerate alcohol.

$$C_4H_8O_2 + KHO = C_2H_6O + KC_2H_3O_2.$$

ETHYL ETHER.

Synonyms : Vinic ether, sulphuric ether, common ether.

$$C_4H_{10}O = \left.\begin{array}{c}C_2H_5\\C_2H_5\end{array}\right\}O.$$

Density .736.
Density of vapor, 37.
Specific gravity of vapor, 2.586.
Boiling point, 35.5°.

To prepare this compound, sulphuric acid is heated with alcohol in a retort, placed in a sand-bath. The ether distills, its vapor being received in a well cooled condenser, provided with a long tube which conducts the uncondensed vapor into a chimney.

The cork adapted to the tubulure of the retort is provided with two openings; in one is fixed a thermometer, through the other a tube passes which furnishes the supply of alcohol. All the connections should close perfectly. When the apparatus is arranged in this manner, pour 700 grams of 85 per cent. or 90 per cent. alcohol into the retort, and add, little by little, 100 grams sulphuric acid of 1.84 sp. gr., then heat. When the thermometer attains 130°, cause the alcohol to flow from the upper vessel at a rate sufficient to keep the temperature between 130° and 140°. The weight of alcohol capable of being transformed into ether is from 13 to 15 times the weight of the mixture first introduced into the retort. The distilled liquid is mixed with 12 parts, to every 100 of its weight, of a solution of soda having a specific gravity of 1.32, and agitated from time to time, during 48 hours.

The ether is decanted by means of a glass siphon, redistilled and four-fifths of the liquid collected. The remainder may serve for a future operation.

This furnishes ordinary ether. To further purify, wash with water, decant and treat for two days with equal parts of quick lime and fused calcium chloride. Williamson has clearly shown that etherification takes place in two stages or successive reactions as follows:

$$C_2H_6O + H_2SO_4 = H_2O + \underbrace{(C_2H_5)HSO_4}_{\text{Ethylsulphuric acid.}}$$

$$(C_2H_5)HSO_4 + C_2H_6O = C_4H_{10}O + H_2SO_4.$$

This explains how a small quantity of sulphuric acid etherizes a large amount of alcohol, since sulphuric acid is constantly regenerated. This is confirmed by the following experiment. Iodide of ethyl is made to react upon potassium alcohol; ether is obtained as indicated by the reaction;

$$C_2H_5I + C_2H_5OK = C_4H_{10}O + KI.$$

Ether is a neutral, volatile liquid, colorless, having a burning taste and a strong agreeable odor. When agitated with water it rises to the surface, but the water dissolves about one ninth of its own weight of the ether. It is miscible with alcohol in all proportions and with wood spirit. Ether is frequently adulterated with the latter substance. Next to alcohol it is the most generally employed solvent for organic substances. It dissolves resin, oils and most compounds rich in carbon and hydrogen.

Bromine, iodine, chloride of gold and corrosive sublimate are soluble in this liquid. It dissolves phosphorus and sulphur in small quantity.

W. Skey (8—Aug. 3, '77,) has shown that contrary to the usual statement in standard works, ether dissolves notable quantities of the alkalies.

At a red heat it is decomposed and furnishes carbon monoxide, water, marsh gas and acetylene.

It is exceedingly inflammable, and burns with a bright flame.

Its extreme volatility, the density of its vapor, its insolubility in water and its great inflammability render its use dangerous, and explosions caused by it are of frequent occurrence. It should never be brought near a fire or light in open vessels. In case ether inflames, it is best, if possible, to at once close the vessel containing it, and thus avoid the more serious consequences ensuing from an explosion. Exposed to the air it experiences a slow combustion as in the case of alcohol, and the same compounds are the result.

Chlorine acts violently upon it; in moderating the action, the whole or a part of the hydrogen may be replaced atom for atom by chlorine.

Uses.—It is used in pharmacy in preparing etherial

tinctures, and as an antispasmodic and stimulant in the well-known Hoffmann's anodyne. Its most important use in medicine is as an anesthetic, than which none is safer or more reliable in efficient hands. It is extensively employed in the laboratory and in photography.

COMPOUND ETHERS

are bodies built up on the type of water, having one half the hydrogen replaced by a hydrocarbide and the other half by a compound radicle containing oxygen, or, in other words, by the radicle of an acid.

$$\text{ACETIC ETHER,} \begin{matrix} (C_2H_5) \\ (C_2H_3O) \end{matrix} \Big\} O.$$

To prepare this ether 8 parts of very concentrated alcohol are distilled with 7 parts of sulphuric acid and 10 parts of anhydrous sodium acetate, which may be replaced by 20 parts of dry lead acetate. The distillate is agitated with a solution of calcium chloride containing milk of lime, decanted, dried over calcium chloride and finally distilled.

Seven parts of water dissolve one part of this body. Alcohol and ether dissolve it in all proportions. It is a solvent for many organic bodies. It is easily decomposed on contact with water. Potassa also effects this decomposition very readily. A prolonged action of ammonia transforms it into acetamide and alcohol.

OXALIC ETHERS.

Oxalic acid, being a bibasic acid, furnishes with alcohol two combinations, one being acid and capable of combining with bases; the other is neutral, $C_6H_{10}O_4$.

Only the latter is of interest. It may be prepared by introducing four parts of 90 per cent. alcohol and four parts of oxalic acid into a retort, adding to this mixture three to six parts of sulphuric acid and then rapidly distilling; the product is washed several times, dried, then redistilled, collecting only the liquid which passes over at 184°. This ether is aromatic, oily, and gradually decomposes in water.

Potassium changes it into carbonic ether.

If oxalic ether is agitated with ammonia, a white powder, *oxamide*, and ethyl alcohol are produced.

$$\begin{matrix}(C_2H_5)_2 \\ (C_2O_2)''\end{matrix}\Big\} O_2 + N_2 \Big\{\begin{matrix}H_2 \\ H_2 \\ H_2\end{matrix} =$$

$$2\left(\begin{matrix}(C_2H_5) \\ H\end{matrix}\Big\} O\right) + N_2 \Big\{\begin{matrix}(C_2O_2)'' \\ H_2 \\ H_2\end{matrix}.$$

Oxamide may be considered as derived from two molecules of ammonia, and belongs to a class of bodies called *diamides*.

It is a white substance, insoluble in cold water and alcohol. Heated with mercuric oxide it is transformed into carbon dioxide and urea. (Williamson.)

Oxalic ether treated with ammonia in solution in alcohol furnishes *oxamic ether.*

In this connection the compounds of the organic radicles with the haloid elements are usually studied: they are not unfrequently denominated ethers of the hydracids. Their type is a molecule of hydrogen, $\left. \begin{array}{c} H \\ H \end{array} \right\}$.

CHLORIDE OF ETHYL OR CHLORHYDRIC ETHER.

$$C_2H_5Cl = \left. \begin{array}{c} C_2H_5 \\ Cl \end{array} \right\}.$$

This body is formed in small quantity when ethylene is made to react upon hydrochloric acid.

To prepare it, alcohol contained in a flask surrounded by cold water, is saturated with hydrochloric acid gas and the mixture then distilled.

$$C_2H_6O + HCl = C_2H_5Cl + H_2O.$$

It is also obtained by pouring into a flask containing 2 parts common salt, a mixture of 1 part alcohol, and 1 part sulphuric acid: it is then gently heated and the ether collected as previously shown.

It is a liquid of an agreeable odor, and very volatile, having a boiling point of 12° and a vapor density of 64°. A red heat decomposes it into ethylene and hydrochloric acid gas. It is combustible and burns with a green, smoky flame; water dissolves the fiftieth part of its volume, alcohol dissolves it completely.

With chlorine it furnishes a complete and regular series of products of substitution which are not identical, but isomeric with the chlorine products of ethene.

Their formulæ are:

$$C_2H_4Cl_2$$
$$C_2H_3Cl_3$$
$$C_2H_2Cl_4$$
$$C_2H\ Cl_5$$
$$C_2\ Cl_6.$$

IODIDE OF ETHYL OR HYDROIODIC ETHER.

$$C_2H_5I = \left.\begin{array}{c}C_2H_5\\I\end{array}\right\}$$

is obtained on causing alcohol to react upon iodide of phosphorus; the action is violent with white phosphorus, considerably less so with red phosphorus.

Six hundred grams of concentrated alcohol are introduced into a retort with 140 grains of amorphous phosphorus, and to this mixture 450 grams of iodine are added. The distilling is carried nearly to dryness. The product, condensed in the receiver, is washed with water containing a little potassa; afterwards with pure water. It is then dried over calcium chloride and again distilled.

Iodide of ethyl is a colorless liquid. Its density is 1.975. It becomes colored on exposure to light, being slightly decomposed; it is again rendered colorless on agitating it with an alkaline solution, which absorbs the acid formed. It burns with a green flame, leaving a residue of iodine. Ammonium compounds in alcoholic, or aqueous solution, furnish ethylamine. This amine can be attacked in its turn by iodide of ethyl and yields diethylamine and oxide of tetrethylammonium. The knowledge of these reactions and their application to other iodides are the basis of a general mode for the preparation of organic bases originated by Hoffmann. Iodide of ethyl, unlike the chloride, is readily decomposed by solutions of silver nitrate, giving a precipitate of silver iodide.

$$C_2H_5I + AgNO_3 = (C_2H_5) NO_3 + AgI.$$

CYANIDE OF ETHYL, OR CYANHYDRIC ETHER.

$$C_3H_5N = \left. \begin{matrix} C_2H_5 \\ C\ N \end{matrix} \right\}.$$

This ether is obtained on distilling in an oil-bath 1 part of potassium cyanide, with 1-5 part of an alkaline sulpho-vinate. To the product, redistilled in a bath of salt-water, nitric acid is slowly added in excess; it is then subjected to another distillation. Finally, it is dried over calcium chloride, and that which passes over from 195° to 200° is collected on redistillation.

Cyanide of ethyl is a colorless liquid of an alliaceous odor, boiling at 97°.

Cyanide of ethyl is decomposed by potassium hydrate; ammonia is produced, and the acid obtained corresponds with a higher homologous alcohol.

$$CN(C_2H_5) + 2H_2O = NH_3 + \underbrace{C_3H_6O_2}_{\text{Propionic acid.}}.$$

M. Meyer observed some years ago, that if cyanide of silver is treated with iodide of ethyl, a liquid is formed, boiling at 82°, of an odor which is not that of ordinary cyanhydric ether. Gautier has shown that this is an isomeric body, and that there are two isomeric series of cyanhydric ethers. Hoffmann has given a distinctive character to these bodies: under the influence of the alkalies they produce a fixed substance, but this is formic acid and not ammonia, and a volatile substance which is a compound ammonia.

$$CN(C_2H_5) + 2H_2O = \underbrace{CH_2O_2}_{\text{Formic acid.}} + \underbrace{\left.\begin{array}{c} H \\ C_2H_5 \\ H \end{array}\right\}N}_{\text{Ethylamine.}}.$$

Organo-metallic Compounds.

Iodide of ethyl attacks the metals and furnishes a class of bodies called *organo-metallic* radicles. None of these bodies are found in nature. They are formed from the iodohydric ethers by the substitution of a metal for the iodine:

$$Zn + 2(C_2H_5I) = (C_2H_5)_2Zn + ZnI_2,$$

$$2Sn + 2(C_2H_5I) = (C_2H_5)_2Sn + SnI_2.$$

Practically these metallic radicles are obtained by various reactions:

1. By the action of the metal upon the iodide, for example;

$$2C_2H_5I + Zn_2 = (C_2H_5)_2Zn + ZnI_2.$$

In certain cases, with tin for instance, the reaction is not as distinct, and there is formed in addition to stannethyl iodide, stannethyl iodides variously condensed.

2d. The metal is treated with another radicle; thus sodium-ethyl is prepared by the action of sodium upon zinc ethyl,

$$(C_2H_5)_2Zn + Na_2 = Zn + 2C_2H_5Na.$$

3d. On decomposing a metalloid compound radicle with a metallic chloride,

$$3ZnCl_2 + 2(C_2H_5)_3P = 3(C_2H_5)Zn + 2PCl_3.$$

4th. Stannethyl is obtained by plunging a plate of zinc into a soluble salt of this radicle: the radicle is precipitated in the form of an oily liquid.

Cacodyl, $As(CH_3)_2$ was the first discovered of this class of bodies. It was obtained by Bunsen on distilling arsenous acid with potassium acetate. The organic radicles combine with metalloids with more or less energy; zinc-ethyl and cacodyl take fire in the air; they also decompose water. The products of oxidation vary with the nature of the compounds employed; zinc-ethyl furnishes the body, C_2H_5ZnO, zinc-ethylate, which, in contact with water, produces alcohol and oxide of zinc. The metals which are less readily oxydized, such as tin, lead and mercury, give oxides which play the parts of bases, and these latter comport themselves like the oxides of the metals they contain. Finally, the radicles formed by the elements, phosphorus, arsenic, and antimony, give, with oxygen, compounds which generally have the character of acids.

Some of the organic derivatives containing phosphorus are very complex. For instance, J. Auanoff (60-'75-493) has obtained a body he denominates, *methyldiethylphosphoniumphenyloxidehydrate!*

To prepare *zinc-ethyl*, we introduce into a flask connected with a condenser inclined in such a manner that the vapors find their way back into the flask, 100 grams iodide of ethyl, 75 grams of zinc, and 6 to 7 grams of an alloy of zinc and sodium, and heat in the water bath until the zinc is dissolved; then the condenser is inclined as usual, and the distilling is effected over a direct fire, collecting the liquid product in a flask filled with dry carbon dioxide. Finally it is again distilled in this gas, and that collected which passes over from 116° to 120°. All the vessels and all the substances should be absolutely dry, and it should always be collected and distilled *in vacuo*, or in carbon dioxide. It is a colorless liquid, whose density is 1.182, boiling at 118°, inflammable on exposure to the air.

With sodium this body furnishes sodium-ethyl, and with chloride of phosphorus or arsenic, it furnishes triethyl phosphine, $P(C_2H_5)_3$, and triethyl arsine, $As(C_2H_5)_3$.

ETHERS. 81

Mercury-methyl, treated with iodine, furnishes a hydrocarbide which has the formula of methyl, CH_3.

Professors Crafts and Friedel (72-[4]19-334) have prepared a large number of compounds of silicon with compound radicles, from which they have deduced valuable theoretical considerations.

MISCELLANEOUS ETHERS.

Formic, butyric, valerianic ether, and other ethers of the fatty series are prepared in the same manner as acetic ether, and have the general properties of this ether. The odor of these ethers is agreeable. Butyric ether has the odor of pine-apple, and valerianic ether that of pears; œnanthylic ether has the aroma of wine, etc. They are used in the manufacture of syrups, flavoring extracts, and for imparting an odor to liquors.

If the difference between the points of ebullition of these ethers is examined it will be seen that the addition of the elements CH_2 causes an elevation of about 20° in the point of ebullition. Kopp has shown that this fact is a general one and applies to the alcohols, and acids of the same series, and to the homologous bodies in general.

	Point of ebullition.	Difference.
Formic ether,	55°	
Acetic "	74°	19°
Propionic "	95°	21°
Butyric "	119°	24°
Valerianic "	133°	14°

The boiling point of one of these bodies may accordingly be predicted, if that of one of its homologous substances is known. There is a certain close relation between the point of ebullition of an ether and that of the acid whose radicle it contains:

	Point of ebullition.	Difference.
Formic acid,	105° ⎱	
" ether,	55° ⎰	50°
Acetic acid	118° ⎱	
" ether,	74° ⎰	44°
Propionic acid,	140° ⎱	
" ether,	95° ⎰	45°
Butyric acid,	163° ⎱	
" ether,	119° ⎰	44°

The solubility in water of the ether formed by homologous acids varies with the molecular weight; thus formic ether is quite soluble, acetic ether is less soluble, butyric ether is but slightly so, and valerianic ether, which follows it, is nearly insoluble.

MERCAPTANS AND THEIR ETHERS.

On substituting sulphur, selenium, or tellurium for oxygen in the alcohols of different atomicity, sulphur, selenium, or tellurium alcohols are obtained, which are designated as mercaptans, selenium mercaptans, and tellurium mercaptans.

Ethers proper correspond to these as to ordinary alcohols. These ethers are derived either by the substitution of an alcohol radicle for the typical hydrogen, as happens with monatomic mercaptans, or by the elimination of H_2S, as is the case with biatomic mercaptans.

One only of each of these two classes will be alluded to here.

Ethyl sulphide, or hydrosulphuric ether, $\quad C_4H_{10}S = \begin{matrix} C_2H_5 \\ C_2H_5 \end{matrix} \Big\} S.$

Ethyl mercaptan, $\quad C_4H_6S = \begin{matrix} C_2H_5 \\ H \end{matrix} \Big\} S.$

To prepare the sulphide a current of ethyl chloride, is passed into an alcoholic solution of potassium sulphide.

The mercaptan is prepared by the action of potassium hydro-sulphide on ethyl sulphide.

In either case potassium chloride is formed.

$$K_2S + 2C_2H_5Cl = C_4H_{10}S + 2KCl$$
$$KHS + C_2H_5Cl = C_2H_6S + KCl.$$

These bodies are afterwards separated by distillation. Like all the sulphur derivatives of alcohol, they have a nauseous odor. The sulphide boils at 91° the mercaptan at 36°.

MIXED ETHERS

containing two different radicles, are obtained by acting, for instance, with ethyl iodide upon potassium methylate, thus:

$$C_2H_5I + \left.\begin{matrix}CH_3\\K\end{matrix}\right\}O = KI + \left.\begin{matrix}CH_3\\C_2H_5\end{matrix}\right\}O,$$

$\underbrace{\hphantom{C_2H_5I}}_{\text{ethyl iodide.}}$ $\underbrace{\hphantom{CH_3/K O}}_{\substack{\text{potassium}\\\text{methylate.}}}$ $\underbrace{\hphantom{KI}}_{\substack{\text{potassium}\\\text{iodide.}}}$ $\underbrace{\hphantom{CH_3/C_2H_5 O}}_{\substack{\text{methyl-ethyl}\\\text{ether.}}}$

or by acting on hydric methyl sulphate $\left.\begin{matrix}CH_3\\H\end{matrix}\right\}SO_4$

with ethyl alcohol. The following is a list of some of the more important mixed ethers of the monatomic series;

TABLE OF MIXED ETHERS.   BOILING POINT.

Methyl-ethyl ether $C_3H_8O = \left.\begin{matrix}CH_3\\C_2H_5\end{matrix}\right\}O$ $+11°$

Methyl-amyl ether $C_6H_{14}O = \left.\begin{matrix}CH_3\\C_5H_{11}\end{matrix}\right\}O$ $92°$

Ethyl-butyl ether $C_6H_{14}O = \left.\begin{matrix}C_2H_5\\C_4H_9\end{matrix}\right\}O$ $80°$

Ethyl-amyl ether $C_7H_{16}O = \left.\begin{matrix}C_2H_5\\C_5H_{11}\end{matrix}\right\}O$ $112°$

Ethyl-hexyl ether $C_8H_{18}O = \left.\begin{matrix}C_2H_5\\C_6H_{13}\end{matrix}\right\}O$ $132°.$

ALDEHYDS.

The following are the principal aldehyds, arranged in series:

$C_nH_{2n}O.$

Formic aldehyd	CH_2O
Ethylic aldehyd	C_2H_4O
Propylic aldehyd	C_3H_6O
Butylic aldehyd	C_4H_8O
Valeric aldehyd	$C_5H_{10}O$
Œnanthylic aldehyd	$C_7H_{14}O$
Caprylic aldehyd	$C_8H_{16}O$
Caproic aldehyd	$C_{10}H_{20}O$
Rutic aldehyd	$C_{11}H_{22}O$
Ethalic aldehyd	$C_{16}H_{32}O$

$C_nH_{2n-2}O.$

Allylic aldehyd (*acrolein*) - C_3H_4O

$C_nH_{2n-4}O.$

Campholic aldehyd (*camphor*) $C_{10}H_{16}O$

$C_nH_{2n-8}O$.

Benzoic aldehyd (*oil of bitter almonds*) C_7H_6O
Toluic aldehyd - - - - C_8H_8O
Cuminic aldehyd - - - - $C_{10}H_{12}O$
Sycocerylic aldehyd - - - $C_{13}H_{22}O$

$C_nH_{2n-10}O$.

Cinnamic aldehyd (*oil of cinnamon*) - C_9H_8O.

Aldehyds may be regarded as bodies built upon the type of one or more molecules of hydrogen, in which one half the hydrogen atoms are replaced by one or more molecules of an oxidized carbohydride.

The formation of aldehyd, (*alcohol dehyd*rogenated), may be illustrated by the following equation:

$$C_2H_6O - H_2 = C_2H_4O$$
$$\underbrace{\qquad\qquad}_{\text{Ethyl alcohol.}} \quad \underbrace{\qquad\qquad}_{\text{Ethyl aldehyd.}}$$

Aldehyds are obtained by the oxydation of alcohols, but they are only the first products of oxydation. They are capable of combining with an additional molecule of oxygen, forming acids; hence the aldehyds are intermediate between alcohols and acids.

ORDINARY ALDEHYD.

$$C_2H_4O = \left. \begin{array}{l} C_2H_3O \\ H \end{array} \right\}.$$

This substance is formed by the slow oxydation of alcohol.

ALDEHYDS.

Alcohol is treated with a mixture of manganese binoxide, or of potassium bichromate, and sulphuric acid, and distilled, care being taken to keep the receiver well cooled. Besides aldehyd, acetyl, acetic ether, acetic acid and water are formed. The product is again distilled, care being taken to collect only that portion which passes over above 60°. This liquid is mixed with ether, and, when cool, a stream of dry ammonia gas is caused to pass through the solution. Crystals of ammonium aldehyd are formed, $C_2H_3(NH_4)O$, which are decomposed by dilute sulphuric acid. The mixture is then distilled.

Aldehyd is a colorless, very volatile liquid. It is soluble in water, alcohol and ether, and possesses a strong, somewhat stifling odor.

The salient property of aldehyd is its avidity for oxygen. If a few drops are poured into water the latter becomes acid; it is therefore a valuable reducing agent.

If aldehyd, or ammonium aldehyd, $\left.\begin{array}{c} C_2H_3O \\ NH_4 \end{array}\right\}$, is poured into an ammoniacal solution of silver nitrate, on slightly elevating the temperature, metallic silver is deposited. This silver adheres to the sides of the tube, and covers it with a mirror-like coating. This property is the basis of a process of silvering glass globes and other hollow articles of glass.

Aldehyd is attacked by chlorine and bromine, and furnishes, by substitution, various products, of which

CHLORAL C_2HCl_3O, is the most important. *Hy-*

drate of chloral, or $C_2HCl_3O + H_2O$, has been prepared now for several years in very large quantities, for medicinal purposes. Its name is derived from *chlor*ine *al*cohol.

Absolute alcohol is saturated, first cold, then hot, with dry chlorine. The liquid obtained is mixed with its volume of concentrated sulphuric acid. The supernatant liquid is decanted, and distilled in an earthern retort, with one-fourth its weight of sulphuric acid. The anhydrous chloral obtained is re-distilled twice with calcium carbonate and 7 to 8 per cent. of water. The hydrate is then obtained in handsome crystals, $C_2HCl_3O + H_2O$, soluble in water. It has been known for some time that this body is decomposed in presence of alkalies or alkaline carbonates, into chloroform and formic acid,

$$C_2HCl_3O + H_2O + KHO = KCHO_2 + CHCl_3 + H_2O.$$

Potassium formiate.   Chloroform.

The question appeared pertinent whether a similar transformation would be effected in the human body, under the action of the alkaline fluids there present, notably those of the blood, and thus develop chloroform.

Liebreich was the first to administer chloral, and he at once obtained the anesthetic effects of chloroform. His experiments were repeated in different countries, and hydrate of chloral soon came into general use as a hyponotic.

Chloral hydrate for medical use must be crystalline and possess the following properties: it should be colorless, transparent, and have an aromatic odor, a caustic taste, readily soluble in water without furnishing drops of oil, also soluble in alcohol, ether, naphtha, benzol, and carbon bisulphide; it should fuse at 56° to 58°, solidify at about 15°, boil and volatilize completely at 95°. With caustic potassa it should furnish chloroform, and with sulphuric acid, chloral, without becoming brown. Its aqueous solution should be neutral and not produce any turbidity with silver nitrate and nitric acid. Exposed to the air it should not become moist. According to recent investigations by Liebreich, (60-69-673) chloral produces the opposite physiological effects of strychnine, hence, these bodies may be used as antidotes one for the other.

The remaining aldehyds are not sufficiently important for a work of this scope. Camphor has already been considered in connection with turpentine.

ORGANIC ACIDS.

ACIDS CONTAINING TWO ATOMS OF OXYGEN.

FATTY ACID SERIES.

$$C_nH_{2n}O_2.$$

Formic	acid,	$C H_2 O_2$
Acetic	"	$C_2 H_4 O_2$
Propionic	"	$C_3 H_6 O_2$
Butyric	"	$C_4 H_8 O_2$
Valeric	"	$C_5 H_{10} O_2$
Caproic	"	$C_6 H_{12} O_2$
Œnanthylic	"	$C_7 H_{14} O_2$
Caprylic	"	$C_8 H_{16} O_2$
Pelargonic	"	$C_9 H_{18} O_2$
Capric	"	$C_{10} H_{20} O_2$
Lauric	"	$C_{12} H_{24} O_2$
Coccinic	"	$C_{13} H_{26} O_2$
Myristic	"	$C_{14} H_{28} O_2$
Palmitic	"	$C_{16} H_{32} O_2$
Margaric	"	$C_{17} H_{34} O_2$
Stearic	"	$C_{18} H_{36} O_2$
Arachidic	"	$C_{20} H_{40} O_2$
Cerotic	"	$C_{27} H_{54} O_2$
Melissic	"	$C_{30} H_{60} O_2.$

ORGANIC ACIDS.

$$C_nH_{2n-2}O_2.$$

Acrylic acid	$C_3H_4O_2$
Crotonic "	$C_4H_6O_2$
Angelic "	$C_5H_8O_2$
Pyroterebic "	$C_6H_{10}O_2$
Campholic "	$C_{10}H_{18}O_2$
Moringic "	$C_{15}H_{28}O_2$
Physetoleic "	$C_{16}H_{30}O_2$
Oleic "	$C_{18}H_{34}O_2$
Doeglic "	$C_{19}H_{36}O_2$
Erucic "	$C_{22}H_{42}O_2.$

$$C_nH_{2n-4}O_2.$$

Sorbic acid	$C_6H_8O_2$
Camphic "	$C_{10}H_{16}O_2.$

AROMATIC ACID SERIES.

$$C_nH_{2n-8}O_2.$$

Benzoic acid	$C_7H_6O_2$
Toluic "	$C_8H_8O_2$
Xylic "	$C_9H_{10}O_2$
Cumic "	$C_{10}H_{12}O_2$
Alpha-cymic acid	$C_{11}H_{14}O_2.$

$$C_nH_{2n-10}O_2.$$

Cinnamic acid	$C_9H_8O_2$
Pinic "	$C_{20}H_{30}O_2.$

ACIDS CONTAINING THREE ATOMS OF OXYGEN.

$$C_nH_{2n}O_3.$$

Carbonic	acid	CH_2O_3
Glycolic	"	$C_2H_4O_3$
Lactic	"	$C_3H_6O_3$
Oxybutyric	"	$C_4H_8O_3$
Oxyvaleric	"	$C_5H_{10}O_3$
Leucic	"	$C_6H_{12}O_3$
Œnanthic	"	$C_{14}H_{28}O_3$

$$C_nH_{2n-2}O_3.$$

Pyruvic	acid	$C_3H_4O_3$
Scammonic	"	$C_{15}H_{28}O_3$
Ricinoleic	"	$C_{18}H_{34}O_3$

$$C_nH_{2n-4}O_3.$$

Guaiacic	acid	$C_6H_8O_3$
Lichenstearic	"	$C_9H_{14}O_3$

$$C_nH_{2n-6}O_3.$$

Pyromeconic acid		$C_5H_4O_3$

$$C_nH_{2n-8}O_3.$$

Salicylic	acid	$C_7H_6O_3$
Anisic	"	$C_8H_8O_3$
Phloretic	"	$C_9H_{10}O_3$
Oxycuminic	"	$C_{10}H_{12}O_3$
Thymotic	"	$C_{11}H_{14}O_3$

ORGANIC ACIDS

$$C_nH_{2n-10}O_3.$$

Coumaric acid - - - $C_9H_8O_3$.

ACIDS CONTAINING FOUR ATOMS OF OXYGEN.

$$C_nH_{2n}O_4.$$

Glyceric acid　　　　　$C_3H_6O_4$.

$$C_nH_{2n-2}O_4.$$

Oxalic	acid	$C_2H_2O_4$
Malonic	"	$C_3H_4O_4$
Succinic	"	$C_4H_6O_4$
Pyrotartaric	"	$C_5H_8O_4$
Adipic	"	$C_6H_{10}O_4$
Pimelic	"	$C_7H_{12}O_4$
Suberic	"	$C_8H_{14}O_4$
Anchoic	"	$C_9H_{16}O_4$
Sebic	"	$C_{10}H_{18}O_4$
Roccellic	"	$C_{17}H_{32}O_4$.

$$C_nH_{2n-4}O_4.$$

Fumaric	acid	$C_4H_4O_4$
Citraconic	"	$C_5H_6O_4$
Terebic	"	$C_7H_{10}O_4$
Camphoric	"	$C_{10}H_{16}O_4$
Lithofellic	"	$C_{20}H_{36}O_4$.

$$C_nH_{2n-6}O_4.$$

Mellitic	acid	$C_4H_2O_4$
Terechrysic	"	$C_6H_6O_4.$

$$C_nH_{2n-8}O_4.$$

Veratric acid	$C_9H_{10}O_4.$

$$C_nH_{2n-10}O_4.$$

Phtalic	acid	$C_8H_6O_4$
Insolinic	"	$C_9H_8O_4$
Choloidic	"	$C_{24}H_{38}O_4.$

$$C_nH_{2n-14}O_4.$$

Oxynaphthalic acid	$C_{10}H_6O_4$
Piperic "	$C_{12}H_{10}O_4.$

ACIDS CONTAINING 5, 6, 7 AND 8 ATOMS OF OXYGEN.

$$C_nH_{2n-2}O_5.$$

Tartronic acid		$C_3H_4O_5$
Malic	"	$C_4H_6O_5.$

$$C_nH_{2n-4}O_5.$$

Mesoxalic acid	$C_3H_2O_5.$

$$C_nH_{2n-6}O_5.$$

ORGANIC ACIDS.

Cholesteric acid $\qquad$ $C_8H_{10}O_5$.

$$C_nH_{2n-8}O_5.$$

Croconic	acid	$C_5H_2O_5$
Comenic	"	$C_6H_4O_5$
Gallic	"	$C_7H_6O_5$
Cholalic	"	$C_{24}H_{40}O_5$.

$$C_nH_{2n-2}O_6.$$

Tartaric acid	$C_4H_6O_6$
Quinic "	$C_7H_{12}O_6$.

$$C_nH_{2n-4}O_6.$$

Carballylic acid $\qquad$ $C_6H_8O_6$.

$$C_nH_{2n-6}O_6.$$

Aconitic acid $\qquad$ $C_6H_6O_6$.

$$C_nH_{2n-10}O_6.$$

Chelidonic acid $\qquad$ $C_7H_4O_6$

$$C_nH_{2n-10}O_7.$$

Meconic	acid	$C_7H_4O_7$
Citric	acid	$C_6H_8O_7$
Mucic	"	$C_6H_{10}O_8$.

Organic acids are bodies built upon the type of one or more molecules of water, having one half the hydrogen replaced by an organic compound radicle con-

taining oxygen. There are some acids whose composition is not definitely fixed. We shall first examine the monatomic acids, and study the other series in the order of their atomicity.

The organic acids possess the general properties of the mineral acids. Many among them, like acetic acid, have a very decided action upon litmus. Generally, they are solid and crystallizable; however, formic, propionic, butyric acids, etc., are liquid. Acids whose molecules are comparatively simple, are ordinarily soluble in water—the others are little, or not at all, soluble in this solvent. The monobasic acids are volatile, at least where their molecules are not very complex. The polybasic acids are decomposed by heat. Their salts are ordinarily crystallizable.

METHODS OF PREPARATION.

I. The acids of the so-called fatty series are obtained by the oxidation of the corresponding alcohol, or aldehyd, which latter is the first product of oxidation of the respective alcohol.

$$C_2H_6O + O = \underbrace{C_2H_4O}_{\text{Acetic aldehyd.}} + H_2O.$$

$$C_2H_4O + O = \underbrace{C_2H_4O_2}_{\text{Acetic acid.}}.$$

II. These acids are also produced by the action of alkalies upon the cyanide of the radicle appertaining to the homologous inferior alcohol.

$$(CH_3)CN + KHO + H_2O = NH_3 + KC_2H_3O_2.$$

<u>Methyl cyanide.</u>   <u>Potassium acetate.</u>

III. Acids are likewise formed by the union of the elements of carbon monoxide and carbon dioxide with hydrogen carbides and water. The remarkable synthesis of formic acid by Berthelot is, according to this method:

$$CO + H_2O = CH_2O_2.$$

Pelouze has shown that heat, carefully applied to polyatomic acids, causes them to part with a certain number of molecules of water, of carbon dioxide, or of both, and furnishes acids more simple and of a lower equivalence, which he designates by the name of *pyro-acids*.

$$2C_4H_6O_6 = C_5H_8O_4 + 2H_2O + 3CO_2.$$

<u>Tartaric acid.</u> <u>Pyro-tartaric acid.</u>

Of all the series of acids, the most numerous and the most important are those of the so-called *fatty series*. We shall presently indicate the methods by which they are obtained.

Their boiling point increases from 15° to 20° with each addition of CH_2 to their molecule. Certain of their salts, those of calcium, for instance, are decomposed by heat, furnishing compounds called *acetones*.

$$Ca(C_2H_3O_2)_2 = CaCO_3 + C_3H_6O.$$

Calcium acetate. Ordinary acetone.

FORMIC ACID.

$$CH_2O_2 = \left. \begin{array}{c} CH\ O \\ H \end{array} \right\} O.$$

Red ants made to pass over moistened blue litmus paper produce red stains. The acid secreted by these insects was first obtained by Gehlen, and has received the name of *formic acid*.

I. Berthelot has obtained it from carbon monoxide by synthesis.

II. It is prepared by distilling a mixture of 10 parts of starch, 30 parts of sulphuric acid, 20 parts of water, and 37 parts of manganese binoxide in a large retort connected with a condenser.

The mass swells considerably, and at first must be heated but gently. The formic acid is distilled over and saturated with lead carbonate. The formiate of lead is caused to crystallize in boiling water, then placed in a retort and decomposed by a current of hydrogen sulphide and thereupon heated; the formic acid is then distilled off.

III. One kilo of glycerine, 150 to 200 grams of water and 1 kilo. of oxalic acid are introduced into a retort and heated for 15 hours at a temperature of about 100°. The oxalic acid is decomposed, but only carbon dioxide is disengaged. Water is added from time to time, and the mixture then distilled until 8 litres have passed over. The glycerine remains unchanged in the retort, and can again be used.

Formic acid is a colorless liquid, of a very acid reaction, a pungent odor and crystallizing at about 0° and boiling at 104°.

It reduces oxide of mercury, furnishing mercury, as a brown powder, also carbon dioxide and water. Its salts are usually soluble, though that of lead is very little soluble in cold water, but quite soluble in boiling water.

On heating with sulphuric acid, carbon monoxide and water are formed.

EXPERIMENT.—Introduce into a test-tube a small quantity of formic acid or a formiate. Add sulphuric acid and heat; a regular liberation of a gas takes place, which may be ignited, producing a blue flame.

$$CH_2O_2 = CO + H_2O.$$

ACETIC ACID.

$$C_2H_4O_2 = \left.\begin{array}{l} C_2H_3O \\ H \end{array}\right\} O.$$

Sp. Gr. 1.08. Density of vapor 30.

Glacial acetic acid melts at 17°; boils at 118°.

This is the acid of vinegar, and of which it forms the essential part. It is found in the juices of many plants and in certain fluids of the body. It is formed by synthesis from methyl, sodium, or potassium formiate, and by the oxidation of acetylene; also by the action of nitric acid upon fatty substances, and by the reaction of potassa upon tartaric, malic and citric acids.

It is further produced:

I. By the oxidation of alcohol in the following way:

Wine in vats, or casks, is placed in a cellar maintained at a temperature of about 30°; every sixth or eighth day several litres of vinegar are withdrawn and replaced by an equal quantity of wine.

Pasteur has established that the oxydation of alcohol is produced by a minute plant, the *Mycoderma aceti*. In fact, acetification commences only when this plant has been formed in the liquid. If its development is interrupted the oxydation stops; it renders the service of taking oxygen from the air and transferring it to the alcohol.

This process is very slow. It may be rendered more rapid by pouring dilute alcohol on beach-wood shavings placed in barrels. The air penetrates through openings made in the lower portion. The alcohol, after having been passed over the shavings four times, will be found sufficiently acetified, if the temperature is maintained at about 25°.

II. DISTILLATION OF WOOD. PYROLIGNEOUS ACID. Wood is distilled in retorts, yielding vapors and gases. The former are condensed by causing them to pass through a condenser; the gases are conducted under the retorts, where they are burned, and the heat utilized in the distillation of the wood.

The condensed liquids are water, acetic acid, wood

ACETIC ACID.

spirit and tar; the greater portion of the tar is mechanically removed and the remaining liquid distilled in a water bath. The wood spirit, which boils at 63° passes into the receiver. The water and acetic acid remaining in the retort are saturated with sodium carbonate, the product is evaporated to dryness and heated from 250° to 350°; this temperature, while not effecting the decomposition of the sodium acetate, is sufficient to carbonize the tarry substance remaining in solution. The mass is thereupon dissolved in water, filtered, and the acetate allowed to crystallize. If it is desired to obtain the acetic acid uncombined, the solution of the salt is distilled with a slight excess of sulphuric acid.

The acetic acid which distils over contains a large amount of water. Normal, or anhydrous acid may be obtained from it by saturating half of the liquid with sodium carbonate, then adding the remainder to this solution; acid sodium acetate is thereby produced, which is evaporated to dryness and distilled with sulphuric acid. This liquid, cooled with ice, gives crystals of normal acetic acid, which can be separated on decanting the liquid, furnishing the so-called *glacial acetic acid*.

Acetic acid is liquid above 17°; below that it crystallizes in handsome plates. It is a strong acid, has a pronounced odor, and is very caustic, producing blisters on the skin. It is soluble in water, alcohol and ether in all proportions. It dissolves resin and camphor, also fibrin and coagulated albumen. On uniting with water it contracts in volume. A red heat destroys it, many products being formed; methane, acetylene, acetone, benzol, naphthalin, etc., also carbon, which remains in the retort.

If a flask containing chlorine gas and a small quantity of acetic acid, is exposed to the sunlight, trichloracetic acid is formed, $\begin{matrix} C_2Cl_3O \\ H \end{matrix} \Big\} O$. This experiment of Dumas served as a basis for the theory of substitution. Le Blanc has also obtained monochloracetic acid $\begin{matrix} C_2H_2ClO \\ H \end{matrix} \Big\} O$. These chlorine products are reduced to the state of acetic acid by reducing agents, such as sodium amalgam in presence of water.

$$(H_2)_3 + C_2HCl_3O_2 = 3HCl + C_2H_4O_2.$$

In the same manner as acetic acid, heated with an excess of a base, furnishes marsh gas, trichlor, acetic acid produces trichlorinated marsh gas, which is chloroform,

$$C_2H_4O_2 + BaO = BaCO_3 + CH_4$$
$$C_2HCl_3O_2 + BaO = BaCO_3 + CHCl_3.$$

Perchloride of phosphorus, in the hands of Gerhardt, has become the means of an important discovery, that of acetic anhydride and in general of the anhydrides of the monobasic acids. If dry sodium acetate (3 parts) is mixed with the perchloride, or better, with oxychloride of phosphorus, (1 part), and then distilled, a chloride is obtained called acetyl chloride,

$$C_2H_3OCl = \left. \begin{array}{c} C_2H_3O \\ Cl \end{array} \right\},$$

acetyl being the radicle of acetic acid. This chloride, subjected to the action of an excess of sodium acetate, is decomposed and furnishes *acetic anhydride*,

$$\left. \begin{array}{c} C_2H_3O \\ C_2H_3O \end{array} \right\} O.$$

(also called acetate of acetyl) or acetic oxide, which boils at 139°. Water destroys it, acetic acid being produced. Chloride of acetyl is an irritating liquid, boiling at about 158°, decomposable by water into acetic and hydrochloric acids.

A derivative of acetic acid of considerable theoretical importance is cyanacetic acid $C_3H_3NO_2 = \left. \begin{array}{c} C_2H_3O \\ CN \end{array} \right\} O,$ a crystalline body forming salts with the metals, which have been studied by T. Menies. On acting with sulphuric acid and zinc on cyanacetic acid, the author [82-67-69] obtained formic and acetic acids and ammonia.

VINEGAR. This name is given to the mixture which is obtained by the acetification of wine, whiskey, infusion of malt, etc. Good acetic vinegar is of an agreeable taste and aroma. Wood vinegar has a very strong disagreeable taste and odor. It is frequently adulterated with sulphuric acid. An addition of $\frac{1}{1000}$ of its weight of this acid is, however, not considered fraudulent, as its presence is regarded necessary to prevent moulding.

A ready method of detecting mineral acids, proposed by M. Witz (77-75-268), is based upon the use of methyl-aniline, which undergoes no change in contact with acetic acid, but promptly changes to a greenish-blue in presence of the least trace of mineral acid.

Vinegar and concentrated acetic acid are employed in medicine as stimulants.

An acetate, or acetic acid, can be recognized by heating it slightly with sulphuric acid and alcohol ; a fragrant odor, characteristic of acetic ether, is observed. Heated with sulphuric acid alone, the acetates liberate a vapor which has the odor of vinegar.

The following reaction permits of the detection of mere traces of acetic acid; it is saturated with potassium carbonate and heated with arsenous oxide in a test tube; fumes and a nauseating odor are given off.

The author finds that one of the simplest tests for acetic acid, is to direct a fine, yet powerful stream of water into a test-tube, containing a few drops of the liquid to be tested. The very fine, white effervescence resulting is entirely characteristic of this acid, none of the other ordinary acids producing the same effect.

Alcohol should not be present, as it causes a similar effervescence. If the acetic acid is combined it should be set free with a strong mineral acid. By this test, perhaps more physical than chemical, acetic acid, diluted with 1000 parts of water, can be readily recognized, and with practice, one part in 1500.

ACETATES.

Acetic acid is monobasic; there are, however, alkaline biacetates and some basic acetates of copper and lead.

POTASSIUM ACETATE.

$$KC_2H_3O_2 = \left.\begin{array}{c} C_2H_3O \\ K \end{array}\right\} O.$$

This salt, distilled with its weight of arsenous oxide, furnishes a very inflammable liquid, formerly called the "liquor of Cadet," and in which Bunsen has found a radicle spontaneously inflammable, *cacodyl*, $C_4H_{12}As_2$.

Potassium acetate forms, as well as sodium acetate, an acid acetate when treated with acetic acid. It is a very deliquescent salt, difficultly crystallizable.

AMMONIUM ACETATE,

$$NH_4C_2H_3O_2,$$

Is prepared by saturating ammonium carbonate with acetic acid. Its solution constitutes the *spirit of Mindererus*; treated with phosphoric oxide it forms cyanide of methyl. There is also an acid salt, $NH_4C_2H_3O_2.C_2H_4O_2$. In compounds of this character, acetic acid must be considered as acting the same part as the water of crystallization in salts.

SODIUM ACETATE.

$$NaC_2H_3O_2 + 3H_2O.$$

This is used in preparing marsh gas and concentrated acetic acid. It is recommended by Tommase (52-72-23), as a solvent for plumbic iodide, of which two grams are readily dissolved in 0.5 c. c. of a strong solution of sodium acetate.

CALCIUM ACETATE.

$$Ca(C_2H_3O_2)_2.$$

This salt, subjected to distillation, furnishes a liquid containing a large proportion of *acetone* C_3H_6O.

ALUMINUM ACETATE.

$$Al(C_2H_3O_2)_3.$$

This body is employed at present by dyers, as a mordant. It is prepared by causing aluminum sulphate to react upon lead acetate. Lead sulphate, which is insoluble, is separated on filtering the liquid.

FERRIC ACETATE.

This salt (*pyrolignite*) has been, and is still, somewhat employed for the preservation of wood.

ACETATES.

COPPER ACETATES.

Normal acetate $Cu(C_2H_3O_2)_2$ is called *verditer*. It forms beautiful green crystals (*crystals of Venus*), which, subjected to distillation, furnish acetic acid mixed with acetone. During this operation, a white sublimate is formed, which deposits in the neck of the retort. This latter is cuprous acetate, and is carried over into the receiver, oxydizes, and changes into cupric acetate, which colors the distillate blue. There remains in the retort, after this decomposition, very finely divided copper which takes fire when slightly heated in the air. Solutions of this acetate reduce the salts of the oxide, CuO, and serve to prepare the suboxide, Cu_2O.

A basic acetate, designated by the name of *verdigris*, is obtained by exposing to the air sheets of copper moistened with vinegar, or surrounded by the *marc* of grapes. The metal becomes covered with a greenish incrustation whose formula is,

$$Cu(C_2H_3O_2)_2, CuO + 6H_2O.$$

LEAD ACETATE.

The normal acetate $Pb(C_2H_3O_2)_2$ is prepared by treating litharge with acetic acid in slight excess. This salt, known by the name of *sugar of lead*, crystallizes in oblique rhombic prisms, soluble in two parts of water and eight parts of 95 per cent. alcohol. It has a sweet taste, and is very poisonous. It is employed as a reagent, also to prepare aluminum acetate and lead chromate.

In digesting acetic acid with an excess of litharge, it furnishes a hexabasic acetate of lead. If ten parts of normal acetate, with seven parts of litharge are taken and this mixture digested with 30 parts of water, there are formed minute needles of a tribasic salt $Pb(C_2H_3O_2)_2$, $PbO2, H_2O$. Finally this salt, dissolved in normal acetate, gives a sesquibasic acetate, which is deposited in crystals, $2(Pb2C_2H_3O_2), PbO, H_2O$.

GOULARD'S EXTRACT is a solution containing a mixture of normal and of sesquibasic acetate of lead, which is prepared by boiling 30 parts of water, 7 parts of litharge and 6 parts of normal acetate of lead.

BUTYRIC ACID.

$$C_4H_8O_2 = \left.\begin{matrix} C_4H_7O \\ H \end{matrix}\right\} O.$$

It is usually prepared as follows: a mixture of 10 parts of sugar, 1 part of white cheese, 10 parts of chalk, and some water, is maintained at a temperature of 30° to 35°. First, lactate of lime is formed, which causes the mass to thicken, then that salt changes into butyrate, disengaging hydrogen and carbon dioxide. When the mixture has become clear, the liquor is evaporated and the butyrate separated with a skimmer. This salt is decomposed by concentrated hydrochloric acid which separates the butyric acid in the form of an oil, which is distilled off. It boils at 163°. It is of a fetid odor, and soluble in water, alcohol and ether.

VALERIC ACID.

Valerianic, or Valeric Acid $C_5H_{10}O_2 = \left. \begin{array}{c} C_5H_9O \\ H \end{array} \right\} O.$

It can be obtained by oxydizing amylic alcohol by a mixture of potassium bichromate and sulphuric acid, or by distilling valerian root with water acidulated with sulphuric acid. The best method is to boil porpoise oil with water and lime. The oil saponifies and the valerianate of calcium alone is dissolved. This liquid is concentrated and hydrochloric acid added in excess. The valerianic acid separates out in the form of an oil which is distilled, and that portion collected which passes over at 175°.

Pierre and Puchot have lately devised a process for preparing valeric acid from amyl alcohol. (3–[3]5–40.)

BENZOIC ACID, $C_7H_6O_2$.

Density, 61.
Density of its vapor compared with air, 4.27.
Melts at 120°; boils at 250°.

It is obtained by a dry, as also by a wet process. To prepare it by the former method, equal weights of sand and gum benzoin are placed in an earthen vessel, the mixture covered with a sheet of filter paper, which is pasted down round the edge, and a long cone of white cardboard placed over the whole. The earthen vessel is then heated over a slow fire for two hours, and when cool the cone is removed. The benzoic acid is found to have condensed on the interior of the cone in handsome blades, or needles.

It is obtained in the wet way, by pulverizing gum benzoin, mixing it with half its weight of lime, and boiling for half an hour in a cast-iron kettle, with six times its weight of water, care being taken to agitate the mixture. It is thrown upon a piece of linen and the residue treated twice with water. The liquids are reduced in volume to two-thirds that of the water used during the first treatment, then saturated with hydrochloric acid. The benzoic acid separates out, and is recrystallized from a solution in boiling water.

It is also procured from the urine of herbivorous animals. This secretion, evaporated to a small bulk and treated with hydrochloric acid, yields a deposit of hippuric acid, which, on being heated with dilute sulphuric acid, is transformed into benzoic acid.

Benzoic acid is also produced on a large scale from naphthalin.

Benzoic acid crystallizes in lustrous blades, or needles, is little soluble in cold water, quite soluble in boiling water, and still more so in alcohol and ether. On passing its vapors through a tube heated to redness, it is decomposed into benzol and carbon dioxide, $C_7H_6O_2 = C_6H_6 + CO_2$. Chlorine, bromine and nitric acid transform it into substitution products.

$$\text{Chlorbenzoic acid, } C_7H_5ClO.$$
$$\text{Dinitrobenzoic `` } C_7H_4(NO_2)_2O_2.$$

Ammonium benzoate furnishes, on distillation, *benzonitrile* $C_7NH_9O_2 = C_7H_5N + 2H_2O$.

The alkaline benzoates heated with chloride, or

BENZOIC ACID. 111

oxychloride of phosphorus, furnish benzyl chloride, which, submitted to the action of potassium benzoate in excess, gives benzoic anhydride,

$$3(KC_7H_5O_2) + POCl_3 = 3(\underbrace{C_7H_5OCl}_{\text{Chloride of benzyl.}}) + K_3PO_4.$$

$$C_7H_5OCl + KC_7H_5O_2 = \underbrace{C_{14}H_{10}O_3}_{\text{Benzoic anhydride.}} + KCl.$$

The rational formula of benzoic anhydride is,

$$\left.\begin{array}{c} C_7H_5O \\ C_7H_5O \end{array}\right\} O.$$

Calcium benzoate heated to a high temperature furnishes *benzone*,

$$\underbrace{Ca(C_7H_5O_2)_2}_{\text{Calcium benzoate.}} = CaCO_3 + \underbrace{CO(C_6H_5)_2}_{\text{Benzone.}}.$$

Benzoic acid is monobasic, and the benzoates are generally soluble. Benzoic acid taken into the stomach, is transformed into hippuric acid.

Kolbe and von Meyer have observed that benzoic acid has antiseptic power, though less than salicylic acid, (18-[2]12-133).

CINNAMIC ACID. In certain balsams there exists an acid called *cinnamic acid*, whose formula is $C_9H_8O_2$. It exists in the balsams of Peru, benzoin, tolu and in liquid storax. It fuses at 129° and boils at 290°. It has striking features of resemblance to benzoic acid, and is produced like the latter by the oxydation of an aldehyd. This aldehyd is the essence of cinnamon prepared by distilling cinnamon with water.

POLYATOMIC ACIDS.

OXALIC ACID.

$$C_2H_2O_4 = \left.\begin{array}{c} C_2O_2 \\ H_2 \end{array}\right\} O_2.$$

PREPARATION. In the burdock and sorrel is found an acid salt, commonly called *salt of sorrel*, which is a mixture of binoxalate and quadroxalate of potassium. Sodium oxalate is found in several marine plants, calcium oxalate in the roots of the gentian and rhubarb, and in certain lichens. Salt of sorrel is extracted from the burdock (*Prunex*), in Switzerland, and in the Black Forest of Germany, by expressing the plant, clarifying the expressed liquid by boiling with clay, and evaporating; crystals of salt of sorrel are deposited.

The oxalic acid may be obtained free by decomposing a solution of these crystals with lead acetate; the oxalate of lead which precipitates is treated with a suitable quantity of sulphuric acid; the lead is completely precipitated as lead sulphate; this is filtered off, and the liquid evaporated and allowed to crystallize.

At present this acid is chiefly prepared by the action of oxydizing agents upon certain organic substances; the substances best suited for this purpose are those

OXALIC ACID. 113

which contain oxygen and hydrogen in the proportion to form water. One part of starch, or sugar, is boiled with eight parts of nitric acid diluted with ten parts of water, until nitrous vapors cease to be disengaged, and the liquid then evaporated. The crystals of oxalic acid which separate out are freed from the excess of nitric acid, by being several times recrystallized in water. It is also obtained on a large scale by the action, at a high temperature, of potassium or sodium hydrate on saw dust.

Oxalic acid has been obtained synthetically, by Drechel, on passing carbon dioxide over sodium heated to $320°$.

$$2CO_2 + Na_2 = Na_2C_2O_4.$$

PROPERTIES.—Oxalic acid crystallizes in prisms, which effloresce in the air, and which are very soluble in water and alcohol.

It fumes at $98°$; at $170°$ to $180°$ it is partially sublimed, but the greater portion is decomposed into carbon monoxide, carbon dioxide, formic acid and water.

$$2(C_2H_2O_4) = CO + 2CO_2 + CH_2O_2 + H_2O.$$

Chlorine, hypochlorous acid, fuming nitric acid and hydrogen peroxide, convert oxalic acid into carbon dioxide.

Sulphuric acid causes it to split up into carbon monoxide and carbon dioxide, and this reaction is made use of in preparing the former gas.

Oxalic acid is bibasic.

Normal potassium oxalate, $K_2=O_2=C_2O_2$.

Acid potassium oxalate, $KH=O_2=C_2O_2$.

USES.—Oxalic acid is employed in removing ink spots from cloth, and in cleaning copper. It owes these properties to the fact that it forms with iron and copper soluble salts, hence it is also employed in calico-works for removing colors.

Toxic action of oxalic acid. On account of the use of oxalic acid in the arts, and its physical resemblance to certain salts, particularly to magnesium sulphate, poisoning with it has often occurred, either through design or imprudence.

It acts powerfully upon the system. Tardieu mentions the case of a young man, sixteen years of age, who was poisoned by two grams of this substance.

The symptoms observed are similar to those produced by other corrosive agents; great prostration followed by unconsciousness and a persistent numbness in the lower extremities. The blood of the patient becomes abnormally red.

In cases of poisoning, the acid should be removed from the stomach with promptness, and milk of lime, or magnesium, or ferric hydrate administered. Lime is to be preferred, as it forms a salt completely insoluble in vegetable acids.

SUCCINIC ACID.

$$C_4H_6O_4 = \left. \begin{array}{l} C_4H_4O_2 \\ H_2 \end{array} \right\} O_2.$$

This acid is produced by the oxydation of butyric acid, and by subjecting amber, *succinum*, to dry distillation or by the action of iodhydric acid on malic or tartaric acids.

Succinic acid crystallizes in rhomboidal prisms which melt at 180° and boil at about 235°, at a higher temperature they are decomposed into water and succinic anhydride $C_4H_4O_3$. It is soluble in 5 times its weight of cold water, soluble in ether and very soluble in alcohol.

It is used in the artificial preparation of malic and tartaric acids. Succinic acid has been found in the fluid of the hydrocele and of certain hydatids.

MALIC ACID.

$$\left. \begin{array}{l} C_4H_3O_2 \\ H,H_2 \end{array} \right\} O_3.$$

This acid, discovered by Scheele in sour apples, is found in many plants; in the berries of the service-tree, in cherries, raspberries, gooseberries, rhubarb, tobacco, etc. Malic acid is levogyrate, deliquescent and crystallizable; it is soluble in alcohol and fuses at about 100°.

At a temperature above 130°, it is decomposed into

various acids and especially *paramalic acid*, $C_4H_4O_4$, which is identical with the acid of the *fumaria*. It is bibasic like oxalic acid, but triatomic and is distinguished from this acid by not producing a turbidity with calcium compounds.

TARTARIC ACID.

$$\left. \begin{array}{c} C_4H_2O_2 \\ H_2, H_2 \end{array} \right\} O_4.$$

This acid, obtained from wine tartar by Scheele, in 1770, occurs free and combined with potassium in many vegetable products; in the sorrel, berries of the service-tree and tamarind, in the gherkin, potato, Jerusalem artichoke, etc. The grape is the chief original source of this acid.

One method of preparing tartaric acid is to purify crude tartar by dissolving and clarifying with clay, which throws down the coloring matters: then filtering and adding calcium carbonate, which precipitates half of the tartaric acid as a calcium salt.

$$2KHC_4H_4O_6 + CaCO_3 = CaC_4H_4O_6 + K_2C_4H_4O_6 + CO_2 + H_2O$$

Hydro-potassic tartrate.   Calcium carbonate.   Calcium tartrate.   Potassium tartrate.

The solution which contains the potassium tartrate, is filtered and calcium chloride added : the remainder of the tartaric acid is thus precipitated as a tartrate and added to the preceding.

$$K_2C_4H_4O_6 + CaCl_2 = CaC_4H_4O_6 + 2\,KCl.$$

<u>Potassium tartrate</u>   <u>Calcium tartrate.</u>

These precipitates are washed and decomposed with sulphuric acid, the calcium sulphate is filtered off, and the liquid evaporated to the point of crystallization. This acid is also called right tartaric, or *dextroracemic*, as it turns the plane of polarization to the right.

Kistner has obtained from certain tartrates a tartaric acid which is optically inactive. This acid, called *paratartaric* or *racemic acid*, is somewhat less soluble than dextrotartaric acid, while the reverse is the case with its salts. It contains, moreover, one molecule of water of crystallization, but does not crystallize, as does the dextrogyrate acid, in hemihedral crystals.

Levogyrate tartaric acid is prepared by evaporating a solution of racemate of cinchonia; the levogyrate tartrate precipitates while the dextrogyrate remains in solution; or a solution of racemic acid is allowed to stand with a small quantity of calcium phosphate, and a few spores of the *Pencilium glaucum;* fermentation sets in, which destroys the dextroracemic acid.

Dextrotartaric acid crystallizes in beautiful oblique prisms with a rhombic base. Cold water dissolves twice its weight of this acid; alcohol dissolves it with equal facility. It is insoluble in ether.

Tartaric acid melts at about 180°; and furnishes different pyrogenous acids, chiefly:

Tartaric anhydride, or *Tartrelic* acid, $C_4H_4O_5$, and *Pyrotartaric acid*, $C_5H_8O_4$.

Simpson synthesized pyrotartaric acid and Lebedeff has recently (60-75-100) shown that this acid is identical with that obtained by heating tartaric acid.

Tartaric acid does not precipitate calcium salts. It produces a turbidity with lime water, but an excess of acid dissolves it; by these reactions it may be distinguished from malic and oxalic acids.

TARTRATES. Tartaric acid is bibasic. The two tartrates of potassium are :

Normal potassium tartrate, $K_2C_4H_4O_6$
Hydro " " $KC_4H_5O_6$.

This latter salt is obtained by purifying the tartar of wine casks, and is called *cream of tartar*. It is used in the preparation of black flux, white flux, potassium carbonate, and tartaric acid, also largely in baking powders.

ROCHELLE SALT. $KNaC_4H_4O_6 + 4aq$. This salt is a double tartrate of potassium and sodium, which was formerly much used as a purgative. It may be prepared by mixing in a porcelain dish, 3500 grams of water and 1000 grams of cream of tartar, this is brought to boiling and sodium carbonate added as long as effervescence is produced. This solution is then filtered and evaporated until it has a density of 1.38.

The salt crystallizes in regular rhomboidal prisms; it is soluble in $2\frac{1}{2}$ times its weight of water, but insoluble in alcohol.

TARTAR EMETIC. Tartaric acid forms, with bases, a

a class of salts called *emetics*, the type upon which they are formed being that of tartar emetic. The ordinary tartar emetic has been generally assigned the formula $(SbO)'K=O_2=C_4H_4O_4$, in which the monad radicle *stibyl* takes the place of one of the basic hydrogen atoms. It is prepared by boiling for an hour in 100 parts of water, 12 parts of cream of tartar, and 10 parts of antimony oxide. This mixture is then filtered, evaporated and allowed to crystallize. This salt crystallizes in rhombic octahedrons; it has a metallic taste, a slight acidity, and is soluble in 14 parts of cold, and about 2 parts of boiling water.

Crystals of tartar emetic effloresce on exposure to the air.

A strip of tin precipitates the antimony as a brown powder. Tannin, and most astringents, precipitate the antimony, hence tartar emetic should not be administered in connection with this class of bodies. This salt is the most used of the antimony compounds.

FERRO-POTASSIUM TARTRATE.—Cream of tartar is digested with ferrous hydrate for two hours at a temperature of 60°. For every 100 parts of cream of tartar, a quantity of hydrate should be used containing 43 parts of ferrous oxide.

The product is filtered, the liquid received in shallow plates, and kept at a temperature of about 45°; the salt thus crystallizes in brilliant scales of a garnet red color. It dissolves in water, but is insoluble in strong alcohol. Tartaric acid is often adulterated with alum, potassium bisulphate and cream of tartar; these substances may all be detected by means of alcohol, in which they are not soluble.

Tartaric acid is used in making effervescing drinks, and as a *discharge* by calico printers.

Tartaric acid produces the same toxical effects as oxalic acid, though requiring *much larger* doses. The blood of the poisoned person becomes red and very fluid.

CITRIC ACID.

$$C_6H_8O_7 = \begin{matrix} C_6H_4O_3 \\ H,H_3 \end{matrix} \Big\} O_4.$$

This acid is found associated with oxalic and tartaric acids in many plants. It occurs in cherries, currants, raspberries, oranges and lemons.

It is ordinarily extracted from the juice of lemons. This juice is allowed to stand until fermentation commences, then filtered and treated with chalk and milk of lime; an insoluble citrate of calcium is formed, which is decomposed by sulphuric acid; the calcium sulphate is filtered off and the filtrate evaporated and left to crystallize. Citric acid crystallizes in regular rhombic prisms; it is soluble in three fourths its weight of cold water; this solution, in time, becomes covered with mould.

Citric acid is soluble in alcohol and ether. Heated to about 175° it furnishes *aconitic acid*,

$$C_6H_6O_6 = \begin{matrix} C_6H_3O_3 \\ H_3 \end{matrix} \Big\} O_3,$$

losing H_2O on increasing the temperature. Another pyrogenous acid, *itaconic acid* $C_5H_6O_4$ is formed, which, if heated, is transformed into *citraconic acid* isomeric with the last mentioned.

Oxydizing bodies destroy citric acid, carbon dioxide, acetone, etc., being produced. Fused caustic potassa resolves it into acetic and oxalic acids.

$$C_6H_8O_7 + H_2O = \underbrace{C_2H_2O_4}_{\text{Oxalic acid.}} + \underbrace{2C_2H_4O_2}_{\text{Acetic acid.}}.$$

Citric acid is tetratomic and tribasic. It may be distinguished from oxalic and tartaric acids by its action on lime water, which it does not precipitate in the cold, but if boiled with an excess of lime water, a precipitate of basic calcium citrate is obtained.

MAGNESIUM CITRATE.—This salt is prepared by treating magnesium carbonate with a strong solution of citric acid and precipitating this salt with alcohol. It is much used in medicine as a purgative.

CITRATE OF IRON.—Hydrated ferric oxide is dissolved in a luke-warm solution of citric acid, and the liquid evaporated to dryness.

This body varies in its composition: it occurs in brilliant amorphous scales, of a garnet-red color.

AMMONIA CITRATE OF IRON.—One hundred grams citric acid are digested for some time with a quantity of ferric hydrate, representing 53 grams of iron, and 16 to 20 grams of aqua ammonia. The liquid is then filtered and evaporated to the consistency of a syrup,

ORGANIC CHEMISTRY.

and transferred to very shallow vessels which are placed in drying ovens. This substance solidifies in scales, if the temperature at which it is dried is not too high and the layers of liquid are extremely thin.

LACTIC ACID.

$$C_3H_6O_3 = \left. \begin{array}{c} C_3H_4 \\ H,H \end{array} \right\} O^3.$$

This acid was discovered by Scheele, who extracted it from sour milk. It exists in many products after fermentation, as sauerkraut, beet juice, and various vegetables, also nux vomica. It is found in many animal fluids, in the blood and in the fluids which permeate the muscular tissues. It is to this body that the acid reaction of sour milk is due. Lactic acid extracted from flesh forms, with certain bases, salts which differ in solubility, etc., from those formed with ordinary lactic acid, hence this acid is sometimes called *paralactic acid*, also *sarko-lactic acid*, from σαρκος flesh.

Lactic acid may be prepared by dissolving sugar of milk in butter-milk, adding chalk to the mixture, and allowing it to stand for eight or ten days at a temperature of 30° to 35°

The sugar of milk is sometimes replaced by glucose, or cane sugar and fermentation favored by the addition of cheese.

A special ferment (*lactic ferment*) is developed which is transformed into sugar and lactic acid, but the fermentation is arrested as soon as the liquid becomes acid, and it is in order to prevent this acidity that an excess of calcium carbonate or sodium bicarbonate is always maintained.

Wurtz has produced this acid artificially by the action of platinum black on propylglycol.

$$O_2 + \underbrace{C_3H_8O_2}_{\text{Propylglycol.}} = C_3H_6O_3 + H_2O.$$

Lactic acid is a colorless, syrupy liquid; at about 130° it is changed into the anhydride of lactic acid, $C_6H_{10}O_5$, and at about 250° it furnishes a crystalline body called *lactide* whose formula is $C_3H_4O_2$.

Lactic acid posseses the property of dissolving calcium phosphate. The lactates are soluble in water. Lactate of iron, $(C_3H_5O_3)_2Fe$, is employed in medicine.

URIC OR LITHIC ACID, $C_5H_4N_4O_3$.

Discovered in 1776, by Scheele.

This acid exists in human excretions, and in those of the carnivora. In the excretions of herbivora, the uric acid is replaced by hippuric acid. Uric acid is present in normal human urine only in small quantity. The urine of sedentary persons, and of those whose food is very nitrogenous and quite substantial, contains more of this substance than that of individuals who lead an active life, and whose diet is less nourishing. In the latter case the uric acid is oxydized and converted into urea, hence, the proportion of the acid decreases as the quantity of urea increases: whereas calculi of uric acid are frequently formed in persons whose diet is very nourishing, and whose occupation necessitates but little muscular exertion. The excreta of birds contains a large proportion of uric acid, and that of snakes is formed almost exclusively of this body.

This acid may be prepared by boiling a dilute alkaline solution with guano, excreta of the boa constrictor, or uric calculi finely pulverized.

The liquid is filtered and the filtrate supersaturated with hydrochloric acid; the uric acid precipitates in flakes, which become crystalline on standing.

The author having had occasion in 1858 to prepare large quantities of uric acid from guano, found that in order to obtain the purest product, as free from coloring matter as possible, it was preferable to use soddium hydrate as a solvent, and carbon dioxide as a precipitant, the latter in sufficient excess to transform the hydrate into bicarbonate.

Crystals of uric acid are colorless and odorless. They are nearly insoluble in ether and alcohol. About 1500 parts of boiling water are necessary to dissolve one part of the acid.

On distillation uric acid yields urea and other cyanic compounds. Uric acid heated with water and lead dioxide furnishes urea and a substance called *allantoin*, which has been found in the urine of sucking calves. Its formula is $C_4H_6N_4O_3$.

The same derivative of uric acid was obtained by the author in 1858, also parabanic acid, on heating uric acid with manganese dioxide and sulphuric acid. (80−[2]41−218.)

URIC ACID. 125

If 1 part of uric acid be added to 4 times its weight of nitric acid of a specific gravity of 1.45, the solution being kept cool, small crystals of a substance called *alloxan* separate out, whose formula is

$$C_4H_4N_2O_5 + 3H_2O.$$

Woehler and Liebig obtained from this body a number of very interesting derivations, *alloxantin, alloxanic acid, parabanic acid, thionuric acid, dialuric acid,* and finally a magnificent purple crystalline body, *murexide.* A large number of various derivatives have also been obtained by other chemists, especially Bayer. The rich color, *murexide,* is made use of in detecting uric acid. For this purpose, traces of uric acid are heated in a watch glass for a few minutes, with one or two drops of nitric acid; the excess of acid is evaporated, and the dry residue exposed to the vapors of ammonia, when a purple, or very beautiful rose color, will appear.

HIPPURIC ACID.

$$C_9H_9NO_3.$$

The urine of herbivora contains a large percentage of this acid, which also exists in a small quantity in human urine. A frugivorous diet augments the proportion of this body. It is prepared by boiling the fresh urine of the horse (hence the name, from ἵππος, a horse), or better from that of a cow, with milk of lime, which is then filtered and evaporated to one-tenth its volume; this is mixed with a large excess of hydrochloric acid and left to stand 10 or 12 hours. The impure hippuric acid which precipitates is re-dissolved in soda and re-precipitated with hydrochloric acid. Animal charcoal may be added to the saline solution if the brown color still remains. Putrid urine yields only benzoic acid. Dessaignes has prepared this acid artificially by causing zincic glycocol to act on benzoyl chloride.

$$Zn(C_2H_4NO_2)_2 + 2C_7H_5OCl =$$
$$ZnCl_2 + 2C_2H_3[NH(C_7H_5O]O_2.$$

Hippuric acid crystallizes in colorless crystals, which require 600 parts of cold water for their solution, but are very soluble in hot water and alcohol.

It is decomposed at 240°, benzoic and cyanhydric acids being found among the products of distillation. Under the action of oxydizing agents it furnishes benzoic compounds; with nitrous acid it yields benzo-glycolic acid.

ALKALOIDS.

ARTIFICIAL BASES OR ALKALOIDS.

PRIMARY.

$$C_nH_{2n+3}N.$$

Methylamine	CH_5N
Ethylamine	C_2H_7N
Propylamine	C_3H_9N
Butylamine	$C_4H_{11}N$
Amylamine	$C_5H_{13}N$
Caprylamine	$C_8H_{19}N$

$$C_nH_{2n+1}N.$$

Acetylamine	C_2H_5N
Allylamine	C_3H_7N

$$C_nH_{2n-5}N.$$

Phenylamine, *aniline*	C_6H_7N
Toluidine	C_7H_9N
Xylidine	$C_8H_{11}N$
Cumidine	$C_9H_{13}N$

$$C_nH_{2n-7}N.$$

Phtalidamine	C_8H_9N

$$C_nH_{2n-11}N.$$

Naphthalamine — — — $C_{10}H_9N.$

SECONDARY.

Dimethylamine — — — C_2H_7N
Methylethylamine — — C_3H_9N
Diethylamine — — — $C_4H_{11}N.$

TERNARY.

Trimethylamine — — C_3H_9N
Dimethylethylamine — — $C_4H_{11}N$
Methylethylamylamine — $C_6H_9N.$

PHOSPHINES.

Methylphosphine — — CH_5P
Dimethylphosphine — — C_2H_7P
Trimethylphosphine — — $C_3H_9P.$

ARSINES.

Triethylarsine — — $C_6H_{15}As.$

STIBINES.

Triethylstibine — — $C_6H_{15}Sb.$

PRINCIPAL NATURAL ALKALOIDS.

OF THE CINCHONAS.

Quinia, Quinicia and Quinidia $C_{20}H_{24}N_2O_2$
Cinchonia and Cinchonidia $C_{20}H_{24}N_2O$
Aricina - - - $C_{23}H_{26}N_2O_4$.

OF OPIUM.

Morphia - - - $C_{17}H_{19}N O_3$
Codeia - - - - $C_{18}H_{21}N O_3$
Thebaia - - - $C_{19}H_{21}N O_3$
Narcotina - - - $C_{22}H_{23}N O_7$
Papaverine - - - $C_{20}H_{21}N O_4$
Narceia - - - $C_{23}H_{29}N O_9$.

OF THE STRYCHNOS.

Strychnia - - - $C_{21}H_{22}N_2O_2$
Brucia - - - - $C_{23}H_{26}N_2O_4$.

OF THE SOLANACEÆ.

Nicotina - - - $C_{10}H_{14}N_2$
Atropia - - - - $C_{17}H_{23}N O_3$
Hyosciamine - - $C_{17}H_{23}N O_3$
Solania - - - - $C_{43}H_{71}N O_{16}$.

OF THE HEMLOCK.

Conylia - - - - $C_8H_{15}N$.

OF PEPPER.

Piperidine - - - $C_5H_{11}N$.

MISCELLANEOUS.

Aconitina - - - $C_{27}H_{40}NO$
Veratria - - - $C_{32}H_{52}N_2O_3$
Theobromine - - $C_7H_8N_4O_2$
Caffeia - - - $C_8H_{10}N_4O_2$.

The first organic base isolated was morphia, obtained in 1816, by Sertuerner. In 1819, Pelletier and Caventou extracted quinia from cinchona bark, and showed that the very active plants used in pharmacy owed their energy to compounds capable of uniting with the acids, and of forming with them definite crystallizable salts.

From that epoch, the number of organic alkaloids has become very considerably augmented; and methods have been discovered by which many of the alkaloids are prepared artificially. It was Fritsche who, in 1840, obtained the first artificial alkaloid on distilling indigo with potassa; he named it *aniline*. Gerhardt by similar methods prepared *quinoleine*, Cahours *piperidine*, and Chantard *toluidine*.

The distillation of organic matter also furnishes alkaloids. Thus several of them have been obtained from a product of the distillation of bones, the oil of Dippel: also as products of the distillation of various other organic compounds.

A very general method is due to Zinin, which consists in causing a reducing substance to act upon nitrous compounds as nitrobenzol, for example. The nitrous compound is introduced into an alcoholic solution of ammonium sulphide, and the mixture allowed to stand; sulphur is soon deposited, and the hydrogen of the hydrogen sulphide combines with the oxygen of the nitrous compound. Example:

$$\underbrace{C_6H_5NO_2}_{\text{Nitrobenzol.}} + 3H_2S = 2H_2O + 3S + \underbrace{C_6H_7N}_{\text{Aniline.}}$$

For this mode of reduction, as it is not very practical, and is tedious in execution, there is at present substituted the action of iron upon acetic acid, or that of zinc or tin, on hydrochloric acid.

Wurtz has given a very interesting method, which has led to the discovery of alkaloids much resembling ammonia, for that reason called *compound ammonias*. It consists in causing potassa to react upon the cyanic ethers, these bodies being decomposed much like cyanic acid.

Thus methylamine is obtained by the action of potassium hydrate upon **cyanate of methyl**:

$$\underbrace{\left.\begin{matrix}CO\\CH_3\end{matrix}\right\}N}_{\substack{\text{Cyanate}\\\text{of methyl.}}} + 2KHO = \underbrace{K_2CO_3}_{\substack{\text{Potassium}\\\text{carbonate.}}} + \underbrace{\left.\begin{matrix}CH_3\\H\\H\end{matrix}\right\}N}_{\substack{\text{Methyl-}\\\text{amine.}}}$$

Hofmann made known, very shortly after the publication of Wurtz' process, a method for the preparation of the compound ammonias, by which not only a simple equivalent of hydrogen is replaced by the radicles (CH_3), (C_2H_5), etc., but all the hydrogen of the ammonia. Hofmann's method consists in causing ammonia to react upon hydrochloric as well as bromhydric or iodhydric ethers, particularly the latter.

Let us take, as an example, iodide of ethyl in connection with the study of

ETHYLAMINE.

Ten to 15 grams of iodide of ethyl and 50 grams of aqua ammonia are heated in sealed tubes of green glass placed in a water bath. The following reaction occurs:

$$C_2H_5I + NH_3 = C_2H_8NI.$$

When the liquid has become homogeneous it is allowed to cool, then decomposed by a solution of potassium hydrate, the vapors being collected in water, containing hydrochloric acid. The hydrochloric acid solution is evaporated to dryness, and the residue treated with pure alcohol, which dissolves the chlorhydride of ethylamine and leaves in an insoluble state the ammonium chloride derived from the excess of ammonia used. The solution of chlorhydride of ethylamine is evaporated to dryness, and the deliquescent crystals obtained decomposed by potassium hydrate, with the aid of a gentle heat. The volatilized product is condensed in a cooled receiver. In this reaction there is also formed diethylamine, triethylamine and oxide of tetrethylammonium from which the ethylamine is separated by distillation.

It may be obtained more readily by first distilling 1 part potassium cyanate with 2 parts potassium sulphovinate, then by decomposing the cyanic ether obtained with a boiling solution of potassium hydrate contained in a flask connected with a cool receiver.

Ethylamine is a limpid liquid, with a strong odor resembling that of ammonia. It has not been solidified. It boils at 18.7°, and dissolves in water, producing a very caustic solution. Ethylamine is equally soluble in alcohol and ether. It is combustible, burning with a blue flame, yellow at the margin.

It displaces ammonia from its combinations. Its solutions give reactions similar to those of ammonia; for instance, with salts of copper it gives a bluish white precipitate, which is dissolved in an excess producing a deep-blue solution.

It differs from ammonia in the following reaction: ethylamine precipitates alumina from its salts, and the precipitate is soluble in an excess of ethylamine, which is not the case with ammonia.

CLASSIFICATION OF THE ALKALOIDS, OR ORGANIC BASES.

AMINES.—Hofmann has given the names of *primary amines*, or *monamines*, to ethylamine, which we have just studied, and the compound ammonias in which a single atom of hydrogen has been replaced by a radicle.

The same chemist, having prepared ethylamine by the action of ethyl iodide upon ammonia, subsequently succeeded in obtaining diethylamine by similar means.

The reaction is the following:

$$N \begin{Bmatrix} C_2H_5 \\ H \\ H \end{Bmatrix} + C_2H_5I = N \begin{Bmatrix} C_2H_5 \\ C_2H_5, HI. \\ H \end{Bmatrix}$$

This hydroiodide obtained, treated with potassium hydrate or lime, furnishes a second base, which is biethylammonia, or diethylamine;

$$\text{Diethylamine } C_4H_{11}N = N \begin{Bmatrix} C_2H_5 \\ C_2H_5 \\ H \end{Bmatrix}.$$

A similar compound is,

$$\text{Ethylaniline } C_8H_{11}N = N \begin{Bmatrix} C_6H_5 \\ C_2H_5 \\ H \end{Bmatrix}.$$

These bases have been given the name of *secondary amines* or *imides*.

The secondary ammonias are attacked by ethyl iodide and other ethers, and a reaction takes place, identical with that which gives rise to the primary and secondary amines and *tertiary amines*, also called *nitrile* bases, are thus obtained.

AMINES.

Such bodies are:

$$\text{Triethylamine } C_6H_{15}N = N \begin{cases} C_2H_5 \\ C_2H_5 \\ C_2H_5 \end{cases}$$

$$\text{Methylethylphenylamine } C_9H_{13}N = N \begin{cases} CH_3 \\ C_2H_5 \\ C_6H_5 \end{cases}$$

These bases are related to the alcohols in the same manner as the primary amines. Thus diethylamine is derived from the action of 2 molecules of alcohol on 1 molecule of ammonia and the elimination of 2 molecules of water:

$$2(C_2H_6O) + NH_3 - 2H_2O = C_4H_{11}N.$$

In like manner the ternary amines may be considered as derived from 3 molecules of alcohol and 1 molecule of ammonia with the elimination of 3 molecules of water.

There are also bodies built upon the type of two and three condensed molecules of ammonia, and are denominated, respectively, di-amines and tri-amines; as

$$\text{Secondary ethylene diamine } N_2 \begin{cases} (C_2H_4)'' \\ (C_2H_4)'' \\ H_2 \end{cases},$$

$$\text{Ternary ethylene diamine } N_2 \begin{cases} (C_2H_4)'' \\ (C_2H_4)'' \\ (C_2H_4)'' \end{cases}.$$

Triethylamine attacks hydroiodic ether, and there is formed the compound $C_8H_{20}NI = N(C_2H_5)_4I$. This body treated with oxide of silver, furnishes an oxygenated *quaternary* base,

$$C_8H_{20}NI + Ag\,HO = Ag\,I + C_8H_{21}NO.$$

This substance is very caustic, soluble in water and acts as an inorganic alkaline base like potassium hydrate, with which body it is also analagous in composition.

$$\left.\begin{matrix}K\\H\end{matrix}\right\}O \qquad \left.\begin{matrix}(C_2H_5)_4N\\H\end{matrix}\right\}O.$$

AMIDES, ALKALAMIDES.—The amides are bodies built upon the type of ammonia, in which one or more of the hydrogen atoms are replaced by an acid compound radicle: thus,

$$\text{acetamide } N\left\{\begin{matrix}C_2H_3O\\H\\H\end{matrix}\right..$$

There are also mixed combinations of amides and amines, called *alkalamides*, as

$$\text{acetanilide } N\left\{\begin{matrix}C_6H_5\\C_2H_3O\\H\end{matrix}\right..$$

NATURAL ALKALOIDS.

Many of the natural alkaloids appear to possess a composition analogous to that of the compound ammonias. Some are not attacked by iodide of ethyl, and should be classified among the *ammoniums*, bodies having the same relation to the compound ammonias as does ordinary ammonium hydrate to ammonia. Others are acted upon by iodide of ethyl, and, from the number of bases furnished, it may be ascertained whether they belong to the primary, secondary or ternary compound ammonias.

The properties of the natural alkaloids in general, resemble those of the artificial bases or alkaloids. They contain nitrogen; those that do not contain oxygen are ordinarily volatile, while those with oxygen are non-volatile; they are very soluble in alcohol, ether and chloroform.

Certain ones are dissolved by the hydrocarbides, which are now considerably used in the preparation of the alkaloids. Water does not dissolve any of the artificial alkaloids, except those having a very low molecular weight, like ethylamine; this liquid, however, dissolves codeia and narceia quite readily. With the exception of quinia and cinchonia, they turn the plane of a polarized ray of light to the left.

They react like ammonia. or potassa, with vegetable

colors, and furnish, with platinum bichloride, crystallizable double chlorides, little soluble and yellow in color. They combine equally well with auric and mercuric chlorides.

The natural alkaloids have ordinarily a bitter taste. Among their salts the sulphates, nitrates, chlorides and acetates are mostly soluble, while the oxalates, tartrates and tannates are insoluble.

The harmless character of tannic acid, and the insolubility of the compounds formed by it, with the alkaloids, render tannin and astringent vegetable substances generally very efficacious antidotes.

The precipitates they produce are soluble in acid and alkaline liquids.

The alkaloids are partially precipitated from their solutions by potassa, soda and ammonia. Iodine water and solutions of iodine in potassium iodide, precipitate them completely.

According to Schultze, the liquid obtained by adding antimony perchloride to a solution of phosphoric acid, is a re-agent which precipitates most of the organic bases.

A delicate re-agent for the alkaloids is the double iodide potassium and mercury. According to Meyer, the best proportions are 49 grams of potassium iodide and 135 grams of mercury dichloride, to 1 litre of water. It is best to *add the re-agent* to the solution of the alkaloid, which may be neutral, acid, or even feebly alkaline.

It must be borne in mind that the presence of

sugar, tartaric acid and of albumen may mask the reactions of a number of alkaloids.

NICOTINA OR NICOTYLIA.

$$C_{10}H_{14}N_2.$$

Nicotina is obtained from tobacco (*Nicotina tabacum.*) For this purpose a decoction of tobacco is made, and the liquor evaporated to a syrup. The extract is treated with twice its volume of 85 per cent. alcohol, which precipitates the salts present and certain organic substances.

The alcoholic solution is distilled and the residue submitted to a second similar treatment. The alcoholic extract thus obtained, is mixed with a concentrated solution of potassium hydrate, and the nicotina liberated is re-dissolved in ether. This ethereal solution is evaporated in a water bath, and the residue distilled in an oil bath, in an atmosphere of hydrogen.

Nicotina is a colorless liquid when pûre, remaining liquid at $-10°$, boiling at about $245°$, with decomposition. It has the odor of an old pipe. Exposed to the air it becomes brown, then resinous; water, alcohol and ether dissolve it; its solutions are strongly levogyrate.

Nicotina is a powerful base; it fumes when a rod moistened with hydrochloric acid is brought near it; it precipitates the metallic oxides. Nicotina requires two molecules of a monobasic acid for saturation. The chloride, $C_{10}H_{14}N_2 2HCl$, is crystallizable, though deliquescent. The hydrogen it contains is not replaceable by methyl, ethyl, etc. It may be considered as having the rational formula,

$$N_2 \begin{cases} (C_5H_7)''' \\ (C_5H_7)''' \end{cases};$$

$(C_5H_7)'''$ being the compound radicle *nicotyl*.

Proportion of nicotina in different tobaccos:

Havana,	- -	2.0 per ct.
Maryland,	- -	2.3 "
Virginia,	- - -	6.9 "
Lothringen,	- -	8.0 "

(Schloesing.)

POISONING BY TOBACCO OR BY NICOTINA.

The injection of a concentrated decoction of tobacco, causes serious results in a few minutes: intense headache is produced, with nausea and vomiting, violent pain in the abdomen, pallor, and, finally, extreme prostration.

An infusion of tea, unroasted coffee, or any astringent substance (pulverized nut-galls, or oak-bark) are the only antidotes known, and they are far from being wholly reliable.

The pure nicotina is one of the most dangerous poisons. It manifests itself immediately on being taken, since it is entirely soluble in water.

The nervous system is especially affected. Two or three drops suffice to cause death.

CONIA. 141

Two drops introduced into the throat of a dog will almost instantaneously cause the following series of symptoms: respiration becomes difficult, the animal staggers, falls without the power of rising again, throws the head back and, in a few moments, is perfectly paralyzed, and death ensues.

PIPERIDINE.

$C_5H_{11}N.$

There has been obtained from the pepper (*Piper longum*, *Piper nigrum* or *Piper caudatum*), a body crystallizing in colorless prisms called *piperine*, whose formula is $C_{17}H_{19}NO_3$. It is a neutral substance. When distilled with three times its weight of soda-lime it furnishes *piperidine*, a limpid liquid having the taste of pepper, and also its odor, soluble in water and alcohol, boiling at 106°.

This body is alkaline and saturates acids. It contains a single atom of hydrogen replaceable by methyl, ethyl, etc.

CONIA, CONYLIA, OR CONINE.

$C_8H_{15}N.$

This body is obtained from hemlock (*Conium maculatum*); the crushed seeds are distilled in a large glass retort, with a solution of potassa, or soda, whereupon an alkaline distillate is obtained. The distilled product is treated with a mixture of two parts of alcohol and one part of ether, which dissolves the sulphate of conia and leaves the insoluble sulphate of ammonium. The ethereal alcohol is separated by distillation, potassa is added to the residue, and the mixture distilled. Water and conia pass over; the latter is dehydrated with potassa, and rectified *in vacuo,* or in a current of hydrogen gas.

Conia is a colorless, oily liquid; emitting an odor of hemlock. Water dissolves it but little, and this better when cold than warm. It is very soluble in alcohol and ether. It boils at about 210°, yet emits vapors even when cold, for if a glass rod, moistened with hydrochloric acid, is brought near it, white fumes are produced. It is a monacidic base, very alkaline, and forms crystallizable salts. One of its atoms of hydrogen is replaceable by ethyl or methyl.

This base is very poisonous. According to Christiason, ten centigrams would suffice to cause death. It is classified among the narcotics; its action is characterized particularly by its effect on the organs of respiration and the left ventricle of the heart.

ALKALOIDS OF THE PAPAVERACEÆ.

The poppy-seed capsules (*Papaver somniferum*) yield, on incision, a milky sap, which dries up in a day or two; this sap, when solidified, constitutes *opium*. There are three leading varieties of opium:

I. *Opium of Smyrna* is found in small cakes of 100 to 150 grams, frequently distorted and agglutinated together by reason of their soft nature, and contain 7 to 10 per cent. of water. The surface is brown, but the interior has a fawn color. Sometimes it is found to contain 14 to 15 per cent. of morphia, but in other instances only 5 to 6. Good Smyrna opium should contain not less than 10 per cent.

II. *The opium of Constantinople* is drier than the preceding. It appears in commerce in flattened, irregular cakes, almost always surrounded with poppy-leaves. It contains 5 to 10 per cent. of morphia.

III. *The opium of Egypt* is still dryer; it is rarely enveloped in leaves. Its odor is feeble, and it contains no more than 2 to 7 per cent. of morphia.

Recently, attempts have been made to cultivate the poppy in Europe, especially in France.

Opium contains the alkaloids morphia, codeia, thebaia, papaverine, opianine, narcotine and narceia, an acid combined with these alkaloids called *meconic acid* (from μηκων, a poppy), a crystallized neutral substance called *meconine*, which, according to Berthelot, is a complex alcohol, and finally, various gummy and resinous compounds.

MORPHIA OR MORPHINE.

$$C_{17}H_{19}NO_3, H_2O.$$

PREPARATION. Ten kilos. of opium are treated repeatedly with water, and the liquors evaporated to the consistency of a syrup.

The mass is redissolved in water, filtered, and again evaporated. To the lukewarm liquid are added 1200 grams of anhydrous calcium chloride, dissolved in twice its weight of water. A complex precipitate is formed, containing resins, coloring matters, and sulphate and meconate of calcium, which is thrown upon a filter.

The filtered liquid is evaporated over a water-bath. During the concentration, a fresh quantity of meconate of calcium is separated by filtering, and the liquid evaporated to the consistency of syrup. The liquid is then acidulated with a small quantity of hydrochloric acid, and set aside in a cool place.

At the end of a few days, it contains brown crystals of the double chlorhydrate of morphia and codeia, contaminated with a blackish liquid; these crystals are drained, pressed, and again dissolved in as little boiling water as possible. The chlorhydrate, on cooling, deposits crystals, which are again dissolved in hot water and decolored with animal charcoal. After heating to 80° or 85°, the solution is filtered, and the liquid, on being concentrated, deposits the double chlorhydrate in pure white crystals.

This salt is again dissolved in boiling water, and the hot liquid treated with ammonia; the codeia remains in solution, while the morphia is precipitated. This deposit is thrown upon a filter washed with cold water, dried, and dissolved in boiling alcohol; the morphia separates out in crystals on cooling.

It frequently contains some narcotina, from which it is freed by washing once or twice with ether, or chloroform, which dissolves the narcotina, and does not affect the morphia.

Pure morphia, (from *Morpheus*, in allusion to its narcotic qualities,) crystallizes in regular prisms with a rhombic base, is colorless, soluble in 500 parts of boiling water, scarcely soluble in cold. Forty to forty-five parts of cold 90 per cent. alcohol are required to dissolve one part of morphia; it is insoluble in ether. Solutions of morphia are very bitter.

Morphia is little soluble in ammonia, while it is dissolved very readily by alkaline solutions, and even by lime water.

Under the action of heat, it fuses in its water of crystallization, the latter escaping, and the alkaloid recrystallizes on cooling.

Morphine is an energetic reducing agent, reducing gold and silver salts, setting free the respective metals. It separates the iodine from solutions of iodic acid. If a solution of starch is poured into a test-tube, and a solution of iodic acid and traces of morphia added, the blue color of iodide of starch appears.

If morphia is put into a few drops of a concentrated and slightly acid solution of a ferric salt, a beautiful blue color is produced, which subsequently changes to green.

Morphia, moistened with nitric acid, is colored orange-red, which rapidly changes to yellow.

These four reactions are characteristic of morphia.

If iodine and morphia are mixed in equal proportions and the mixture treated with boiling water, a brown liquid is formed which deposits a reddish-brown powder called *iodomorphia*. Morphia fused with alkalies yields methylamine. (p. 127). It is attacked by ethyl iodide at 100°, a single molecule of ethyl entering into the group.

Morphia forms crystallizable salts, from the solutions of which it is precipitated by the fixed alkalies.

CHLORHYDRATE OF MORPHIA, $C_{17}H_{19}NO_3HCl+3H_2O$. To prepare this salt, 100 parts of pulverized morphia are treated with a little warm water, then hydrochloric acid is added in sufficient quantity to dissolve the alkaloid. The solution is afterwards evaporated in a water bath until it crystallizes.

This salt is soluble in 20 parts of cold water, very soluble in alcohol. It is the salt of morphia most used, and contains 76 per cent. of morphia.

SULPHATE OF MORPHIA, $(C_{17}H_{19}NO_3)_2H_2SO_4+5H_2O$ is prepared like the preceding salt, which it resembles in appearance as well as in properties.

Morphia and its salts are used in very small doses, as in larger doses they are energetic poisons.

CODEIA, $C_{18}H_{21}NO_3,H_2O$.

Discovered in 1832 by Robiquet. This base, whose name is derived from κώδη, poppy head, exists in the ammoniacal solution obtained in the preparation of morphia. On evaporation the ammonia is driven off and the codeia is precipitated by potassa. The codeia is at first precipitated in the form of a sticky mass which soon becomes pulverescent. It is washed with and dissolved in hydrochloric acid. The liquid is then boiled with washed animal charcoal, and the codeia precipitated with potassa.

Codeia is crystalline, very soluble in alcohol and ether. It dissolves in 80 parts of cold and in 20 parts of boiling water.

Codeia is very soluble in ammonia, and nearly insoluble in potassa. With chlorine, bromine and nitric acid it forms products of substitution. With iodine it furnishes ruby-red crystals, whose formula is

$$C_{18}H_{21}NO_3I.$$

Codeia is somewhat used as an anodyne. It is easily distinguished from morphia, since:

I. Codeia is soluble in ether and ammonia.
II. It is insoluble in solutions of potassa.
III. It does not reduce iodic acid or ferric salts.
IV. Nitric acid does not impart to it any color.

Narcotina. $C_{22}H_{23}NO_7$.

Narcotina crystallizes in rhombic prisms. It is almost insoluble in cold water, somewhat soluble in alcohol, quite so in ether. It fuses at 170°, and is decomposed before reaching 200°. Dilute nitric acid transforms it into various products of oxydation, the most important of which are *meconine*, *cotarnine* and *opianic acid* Narcotina unites with acids, but the compounds are decomposed on evaporation.

It is distinguished from morphia in that it does not reduce iodic acid and ferric salts, and from codeia in giving with nitric acid a blood red coloration. This substance is also insoluble in potassa and ammonia. It is not as poisonous as morphia.

THEBAIA.

$$C_{19}H_{21}NO_3.$$

This alkaloid, sometimes called *paramorphia*, is the most poisonous of the bases of opium.

It is crystallizable, insoluble in water, soluble in alcohol and ether. Fuming nitric acid attacks it in the cold, and a yellow liquid is obtained, which becomes brown on contact with alkalies, and which disengages an alkaline vapor. Concentrated sulphuric acid gives it a red hue.

PAPAVERINE.

$$C_{20}H_{21}NO_4.$$

This body is crystallizable, insoluble in water, quite soluble in boiling alcohol and ether. It forms crystalline salts.

Under the action of strong sulphuric acid it assumes a deep blue color, though Hesse and Dragendorff have recently ascertained that when absolutely pure no color is obtained, the ordinary article found in trade not being pure.

NARCEIA.

$$C_{23}H_{29}NO_9.$$

This alkaloid crystallizes in silky needles, insoluble in ether, soluble in alcohol and boiling water, little soluble in cold water. It forms crystallizable salts.

Narceia fuses at 95°, and commences to decompose at about 110°. It is attacked in the cold by concentrated sulphuric acid, a red liquid being produced which rapidly becomes green, especially if slightly heated. The best means of distinguishing narceia is to cause a solution of iodine to act upon the pulverized substance. According to Roussin, the operation is most easily performed with one part of iodine and two parts of potassium iodide dissolved in ten parts of water. A blue color is produced, which disappears on coming in contact with alkalies, or on heating.

PHYSIOLOGICAL ACTION OF OPIUM. NARCOTIC POISONS.

Opium in small doses is a very highly-prized anodyne. Continued use of this substance produces a peculiar state of inebriation, an excited sleep and hallucinations of various sorts.

The bodies of opium-eaters are lean and cadaverous, their eyes are lustrous, their forms bent; their appetite diminishes, and they exist only by increasing the dose of the poison which destroys them. In larger doses it is highly poisonous, and acts in a different manner from that of the poisons already studied. It may be considered as the type of the narcotic poisons. It is not unfrequently used for criminal purposes, and the imprudent administration of laudanum and other solutions of this substance often causes serious effects.

Claude Bernard has made a careful study of the action of the various alkaloids of opium upon the system, and has tabulated their soporific, toxic, and convulsive actions as follows:

Toxic.	Convulsive.	Soporific.
Thebaia,	Thebaia,	Narceia,
Codeia,	Papaverine,	Morphia,
Papaverine,	Narcotina,	Codeia.
Narceia,	Codeia,	
Morphia,	Morphia,	Without action.
Narcotina.	Narceia.	

Those at the head of each column are the most marked in the respective characteristic action.

Subjoined are tabulated the principal chemical characteristics of the opium alkaloids:

	WATER.	ALCOHOL.	ETHER.	AMMONIA.
Morphia.	But little soluble.	Quite soluble.	Almost insoluble.	Nearly insoluble.
Codeia.	Soluble.	Very soluble.	Very soluble.	Soluble.
Narcotina.	Insoluble.	Soluble.	Soluble.	Insoluble.
Thebaia.	Insoluble.	Soluble.	Soluble.	Insoluble.
Papaverine.	Insoluble.	Soluble.	Soluble.	Insoluble.
Narceia.	Slightly sol'ble	Soluble.	Insoluble.	Insoluble.

QUINIA OR QUININE.

$$C_{20}H_{24}N_2O_2, 3H_2O.$$

This alkaloid was discovered in 1820 by Pelletier and Caventou. The following is the modern process by which it is prepared.

Yellow Peruvian bark is carefully pulverized and thoroughly mixed with 30 per cent. of its weight of lime, previously slacked. The mass is then lixiviated three or four times with refined petroleum (petroleum ether) or amylic alcohol, (wood spirit) which dissolves the alkaloids.

POTASSA.	NITRIC ACID.	SULPHURIC ACID.	IODIC ACID.
Soluble.	Orange-red coloration.	Colored violet on heating with dilute acid.	Reduced.
Nearly insoluble.	Orange-red coloration,	Colored violet on heating with dilute acid.	Is not reduced.
Insoluble.	Blood-red coloration.	Yellow coloration.	Is not reduced.
Insoluble.	Yellow coloration.	Red coloration.	
Insoluble.		Dark-blue coloration.	
Insoluble.		Red color, which becomes green.	

The united extracts are agitated with water, acidulated with sulphuric acid, making the liquid only slightly acid.

When the solution is completed, animal charcoal is added, and the liquid brought to boiling, filtered while still hot, and allowed to cool. . The quinia sulphate which is formed, $2(C_{20}H_{24}N_2O_2)$, $H_2SO_4+7aq.$, being but slightly soluble, is deposited on cooling.

After being allowed to stand 24 hours, the sulphate is collected, expressed and redissolved in as small a quantity of water as possible, containing a few drops of sulphuric acid.

The liquid on cooling, deposits crystals, which are dried at 35°. The mother liquors are treated with ammonia, or sodium carbonate, which precipitates a certain quantity of the alkaloid. The precipitate is lightly washed with water, redissolved in dilute sulphuric acid, boiled with washed animal charcoal, and allowed to cool. A second crop of crystals of quinia sulphate is thus obtained. The mother liquor contains cinchonia sulphate. This sulphate is dissolved in 30 times its weight of boiling water, allowed to cool, and a slight excess of ammonia added.

The cinchonia which is precipitated is collected on a filter, and washed with lukewarm water until the filtrate no longer gives with barium chloride a white precipitate insoluble in acids; it is then dried at a temperature of 30° to 40°.

Quinia is white, amorphous and very friable. It

SULPHATES OF QUINIA. 153

may be obtained in a crystalline condition, by adding an excess of ammonia to a dilute solution of quinia sulphate, and allowing the solution to stand.

This crystallized quinia melts at 57°, losing its water of crystallization, solidifies and remelts at 176°. It requires 250 parts of boiling and 460 parts of cold water for its solution.

It dissolves in 2 parts of boiling absolute alcohol, 2 parts of chloroform or 50 to 60 parts of ether. Its solutions are very bitter, levogyrate, and for the most part fluorescent.

Heated on platinum foil, quinia swells up and inflames, leaving a deposit of carbon. Heated with potassa it produces hydrogen and *quinoleine;* (cinchonlein); it also furnishes a brown compound on being triturated with iodine.

Quinia is recognized by the following reactions. It is first saturated with very dilute sulphuric acid and chlorine water; then an excess of ammonia is added, whereupon a green color is obtained.

On adding powdered potassium ferrocyanide before the aqua ammonia a rose coloration is produced, which afterwards becomes dark red.

Quinia has a basic reaction; it forms with acids crystallizable salts from which the alkalies precipitate quinia. It is a base which saturates two molecules of a monobasic acid.

SULPHATES OF QUINIA. Two sulphates of quinia are known; that obtained by the process we have above described, is the neutral sulphate, though generally known as the basic sulphate. Its formula is

$$2C_{20}H_{24}N_2O_2.H_2SO_4+7H_2O.$$

This salt contains 74.3 per cent. of quinia.

It crystallizes in very delicate needles belonging to the clinorhombic system, and which effloresce in dry air. It dissolves in 30 parts of boiling and 740 parts of cold water; also in 60 parts of cold absolute alcohol. It is very nearly insoluble in ether. Its solutions are extremely bitter. It becomes phosphorescent on being heated, and subsequently fuses.

Heated in the air it burns, leaving a carbonaceous residue.

On adding quinia to water acidulated with sulphuric acid, it rapidly dissolves and another sulphate, often called the acid sulphate, is formed, whose formula is.

$$C_{20}H_{24}N_2O_2.H_2SO_4+7H_2O.$$

It is on account of the difficult solubility of the preceding salt, and the great solubility of this latter one, that we cautioned against the employment of an excess of sulphuric acid in the preparation of quinia.

This salt dissolves in 11 parts of water at 12°, and in 9 parts at 18°. Sulphate of quinia, heated to 130° with acidulated water for several hours, is transformed into an isomeric dextrogyrate base called *quinicine*, which is likewise a febrifuge.

Medicinal sulphate of quinia always contains sulphate

of cinchonia, and its presence is not considered fraudulent, even when it contains 3.5 per cent. of the latter substance, as this salt is necessarily produced in the preparation of quinia. Cinchonia appears to be of little therapeutic value, and is often added to sulphate of quinia.

This adulterant is detected by weighing out 0.5 grams of the salt, and adding to it 5 grams of ether. The mixture is agitated and 1.5 grams of concentrated ammonia added. If no cinchonia is present, two liquid layers are obtained; if it is present, a layer of this alkaloid is formed directly above the ammonia. Good commercial sulphate of quinia should give only a very thin layer.

The amount of quinia may be directly determined by decanting and evaporating the ethereal solution, and weighing the residue. This result may be verified by replacing the ether in another determination, by chloroform, which dissolves both bases; the residue obtained by the evaporation of this liquid furnishes the weight of the quinia and cinchonia together.

Sulphate of quinia sometimes contains sulphate of quinidia; this base is precipitated, together with cinchonia, by ether. Its presence may be detected by dissolving one gram of the sulphate in 30 grams of boiling water, and adding to the solution ammonium oxalate. Oxalate of quinidia, which is the only soluble oxalate of these bases, remains in solution, and, on filtering, a bitter liquid will be obtained, in which the quinidia may be precipitated by ammonium hydrate.

In case sulphate of quinia has been adulterated with calcium sulphate, or other inorganic substance, it may be recognized by a residue which will be obtained on heating the sulphate to redness on platinum foil.

Sulphate of quinia should dissolve in 80 per cent. alcohol. If it dissolves in water, but does not dissolve in 56 per cent. to 60 per cent. alcohol, it may be regarded as not pure.

If adulterated with starch, or fatty bodies, a clear solution cannot be obtained, even in very large quantities of water.

Should it contain sugar it will emit an odor of caramel on ignition, and blacken in contact with sulphuric acid.

Quinia sulphate to which salicin, a common adulterant, has been added, is colored red by sulphuric acid.

Quinia sulphate is chiefly employed in cases of intermittent fevers.

CINCHONIA OR CINCHONINE.

$$C_{20}H_{21}N_2O.$$

Cinchonia was discovered by Duncan in 1803, though first recognized as an organic base by Pelletier and Caventou in 1820.

It differs from quinia in containing one atom less of oxygen; it has never been converted into quinia.

It is prepared in the same manner as quinia, but

from the Gray Peruvian Bark. Cinchonia separates out in crystals on the evaporation of the alcohol with which the calcic precipitate is washed.

The crystals of cinchonia are collected, allowed to drain, and the liquid which runs off will furnish additional crystals on being evaporated. To this mother liquor sulphuric acid is added in excess, and the solution slightly evaporated.

The first crystals obtained are sulphate of quinia. which is less soluble than sulphate of cinchonia. When nothing remains but a very concentrated motherliquor, the cinchonia is precipitated by ammonia, and freed from quinia by washing with ether. The quinia dissolves, while the cinchonia remains insoluble.

The latter crystallizes in brilliant colorless crystals, which are insoluble in cold water and ether, soluble in 2,500 parts of boiling water, in 30 parts of boiling 90 per cent. alcohol, and 40 parts of chloroform.

Its solutions are very bitter and dextrogyrate.

Cinchonia melts at about 257°; on heating to a slightly higher temperature in a current of nitrogen, or hydrogen, it is completely sublimed.

With chlorine and bromine, it furnishes dichloride and dibromide of cinchonia. With iodine, a yellow crystalline body is obtained, whose formula is $C_{20}H_{21}N_2OI$.

Heated with fused potassa, it produces *quinoleine*.

Cinchonia has an alkaline reaction. It unites with acids, forming salts which correspond to the salts of quinia, though generally more soluble.

Cinchonia sulphate, heated to about 135°, furnishes the sulphate of an isomeric alkaloid, *cinchonicia* or *cinchonicine*.

Cinchonia is employed as a febrifuge in Holland, and a few other countries, but its action is regarded as inferior to that of quinia.

QUINOIDINE.—*Quinidia* is a base obtained from the last mother-liquor in the preparation of quinia, by precipitation with sodium carbonate. It is often mingled with another alkaloid, *cinchonidia* or *cinchonidine*, and it is this mixture, containing chiefly quinidia, which is called *quinoidine* in commerce.

Quinidia is isomeric with quinia; it melts at 160°. It is difficultly soluble in water, very soluble in boiling alcohol, and slightly soluble in ether. Its solutions are dextrogyrate. Quinidia acts as a febrifuge. With chlorine and ammonia, it gives the same reactions as quinia, and forms corresponding salts.

Quinoidine contains, as we have said, cinchonidia, a substance isomeric with cinchonia. This body is crystalline, fusible at about 150°, almost insoluble in water, slightly soluble in ether and chloroform; boiling alcohol is the best solvent for cinchonidia.

ALKALOIDS OF THE STRYCHNOS.

The two chief alkaloids are strychnia and brucia. Desnoix extracted from the nux vomica another alkaloid, which he named *iyasuria*; but according to Schützenberger, this body is a mixture of several bases.

These alkaloids are extracted from the fruit of the *Strychnos nux vomica*; from St. Ignatius' beans, fruit of the *Strychnos Ignatii*; from the wood of Coulevre, root of the *Strychnos colubrina*; from the upas, the poison of indian arrows, extracted from the *Strychnos tieuté*; from the False Angustura Bark, also from the bark of the *Strychnos nux vomica*, which contains principally brucia.

STRYCHNIA.

$$C_{21}H_{22}N_2O_2.$$

Nux vomica is pulverized and boiled with three successive portions of water containing sulphuric acid, and these decoctions evaporated in a water bath. When the liquid is reduced to a small volume, 125 grams of quicklime slacked to a thin paste are added for each kilo. of nux vomica. The precipitate is collected on a cloth, washed, dried, and treated with 90 per cent. alcohol.

The alcoholic solution is distilled to three-fourths its volume and left to crystallize. The crystals obtained are chiefly strychnia; these are allowed to drain, then dissolved in water containing $\frac{1}{20}$ its weight of nitric acid, and the solution concentrated in a water bath.

The nitrate of brucia remains dissolved and the nitrate of strychnia crystallizes out. These crystals are re-dissolved in water, animal charcoal added, the solution brought to boiling and then filtered.

Ammonia is added to this liquid, the precipitate washed, dried, and dissolved in boiling alcohol, which deposits the alkaloids on cooling.

This method is at present very advantageously supplanted by the process given for the production of quinia, which, briefly stated, consists in treating the substance with lime directly and employing a solvent for the alkaloids, which is insoluble in water, such as petroleum or amylic alcohol.

Strychnia crystallizes in octahedrons or in prisms of the rhombic system; they are colorless, very bitter, and almost insoluble in water or ether, but readily soluble in ordinary alcohol diluted with 75 per cent. of water. Strychnia treated with potassa furnishes a small quantity of quinoleine. Iodide of ethyl produces with this base the compound.

BRUCIA. 161

$$C_{21}H_{22}(C_2H_5)N_2O_2I.$$

Chlorine gas renders even a dilute solution of this alkaloid turbid and the liquid becomes acid; this reaction is characteristic. Bromine also forms derivatives by substitution. Iodine combines directly with the molecule of strychnia.

Strychnia dissolves in strong sulphuric acid; the solution is colorless and becomes dark blue in contact with potassium bichromate or lead dioxide. The color rapidly passes to red and finally to a yellow.

Strychnia is colored yellow by hydrogen nitrate only when it contains brucia, a trace of which is sufficient to produce the change.

Strychnia forms with acids crystallizable salts. The nitrate $C_{21}H_{22}N_2O_2,HNO_3$ crystallizes in fine needles very soluble in hot water.

Strychnia is among the most powerful poisons, 2 to 3 centigrams being sufficient to cause death. There is believed to be no reliable antidote for strychnia though F. M. Peirce claims that small doses of prussic acid are efficient for the purpose. (44-'68-335.)

BRUCIA.

$$C_{23}H_{26}N_2O_4,4H_2O.$$

To obtain this alkaloid the alcoholic liquids from which strychnia has been removed, are saturated with oxalic acid and evaporated. The crystals of oxalate of brucia which are formed, are washed with 95 per cent. alcohol and redissolved in water. The solution is decomposed by lime, the precipitate collected, dried and dissolved in boiling alcohol; brucia then crystallizes out and is purified by two recrystallizations.

Crystals of brucia are large and of the clinorhombic system; they are soluble in alcohol, insoluble in ether, but soluble in 850 parts of cold, or 500 parts of boiling water.

Concentrated sulphuric acid strikes a rose color with brucia which afterwards changes to green. Nitric acid colors it red, and if heated it gives off nitrous ether, methyl alcohol and carbon dioxide.

Brucia is much less poisonous than strychnia.

It may be distinguished from strychnia by its reaction with nitric acid. A red color is produced by brucia, which passes to violet on the addition of stannous chloride. This latter coloration does not take place with morphia. Brucia is also one of the best reagents for nitric acid.

CURARINA.—From the arrows of the Indians living on the shores of the Amazon and Orinoco, a brown resinous matter is collected, from which crystals of a substance have been obtained whose poisonous action is exceedingly rapid. Preyer, to whom we owe this discovery, regards its formula as $C_{10}H_{15}N$, and has named it *curarina*.

The Indians of Dutch Guiana poison their arrows with two other substances no less dangerous: *urari* and *tikunas*. These three substances paralyze the action of the muscles by destroying the motor nerves

(Claude Bernard). It appears that urari, though a fatal poison when introduced into the blood by a wound, may yet be swallowed with impunity.

DRASTIC POISONS.

We shall not describe the preparation of the following alkaloids, on account of their minor importance. The process in general is similar to that by which the preceding ones are prepared: The alkaloid is dissolved in an inorganic acid, precipitated by a base, and redissolved in an appropriate solvent.

The roots of the white hellebore (*Veratrum album*) and its seeds, furnish an alkaloid called *veratria*, $C_{32}H_{52}N_2O_8$. It crystallizes in prisms having a rhombic base. They are very bitter, insoluble in water, soluble in alcohol and ether, and melt at 115°. Veratria is dissolved by strong nitric acid, the solution being violet. Sulphuric acid colors it first yellow, then red.

Three other poisonous bases, *sabadillia*, *colchinia*, and *jervia*, are found associated with veratria in the *Veratrum album*. Jervia, $C_{20}H_{46}N_2O_3 2H_2O$, (Gerhardt and Wills' analysis) is white, crystalline and fusible.

These bodies are very corrosive poisons, producing great irritation of the alimentary canal.

ALKALOIDS OF THE POISONOUS SOLANACEÆ.

The belladonna, *Atropa belladonna*, and the thornapple, *Datura stramonium*, furnish each an alkaloid called, respectively, *atropia* and *daturia*, the formula of which is $C_{17}H_{23}NO_3$.

This substance crystallizes in fine needles, which are fusible at about 90°, and are partially sublimed at about 135°. It is difficultly soluble in water, but very soluble in alcohol and ether.

Heated with an oxydizing agent, such as potassium bichromate, or sulphuric acid, it disengages essence of bitter almonds, easily recognizable by its odor, and crystals of benzoic acid are sublimed. With sulphuric acid a violet color is produced, accompanied by a fragrant odor resembling that of a rose.

Hydrochloric acid furnishes two acids with atropia, *tropic* $C_9H_{10}O_3$, and *atropic* $C_9H_8O_2$.

Cases of poisoning by atropia are rare, but instances in which persons are poisoned by the berries of belladona are of frequent occurrence.

The black henbane, *Hyoscyamus niger*, furnishes silky needles of a substance, *hyosciamine*, which has much resemblance to atropia, but whose action as a poison appears to be less violent.

Its physiological action is on the nerves rather than on the muscles. It causes less dilation of the pupil of the eye, and produces a sombre delirium.

Belladonna and atropia, datura, also henbane and hyosciamine, as well as the poisonous solanaceæ in general, should be classed among the narcotic poisons.

Poisoning produced by belladonna, and by most of the poisonous solanaceæ, is characterized by great dilation of the pupils of the eyes. The patient is also seized with vertigo and strange hallucinations followed by a turbulent delirium and convulsions. The face is congested, respiration difficult, and the skin often breaks out in an eruption similar to that in rubeola (measles).

No antidote is known for these poisons; an infusion of unroasted coffee, tea, or other astringent substances is recommended, but the use of energetic emetics and purgatives is the most efficient method of treatment.

The chemical characters of these alkaloids has not been as yet very fully studied.

Desfosse has extracted from the woody nightshade, *Solanum dulcamara*, from the berries of the felon-wort and from the young sprouts of the potato, *Solanum tuberosum*, a substance called *solanine*, $C_{43}H_{71}NO_{16}$, a highly poisonous alkaloid. On being boiled with acids, it furnishes a stronger base *solanidine* and glucose.

ACONITINA.

Aconitina is extracted from the monk's-hood, *Aconitum napellus*, as a colorless amorphous, bitter powder, soluble in alcohol, slightly soluble in ether, and almost insoluble in water. It fuses at 120°, and is alkaline. It is a very active poison. Planta gives its formula as $C_{30}H_{47}NO_7$ (?).

Duquesnel has extracted from the *Aconitum napellus* a crystalline alkaloid, whose formula is $C_{27}H_{40}NO$.

DIGITALIN.

This substance was long ago obtained in an amorphous condition from the purple fox-glove. In 1871 Nativelle succeeded in obtaining it in a crystalline form. An extract of fox-glove is first prepared, concentrated by distillation and dilluted with 3 times its volume of water.

A precipitate is formed which contains two bodies, *digitalin* and *digitin*. This deposit, washed with boiling alcohol, furnishes crystals composed of these two substances, which are easily separated by chloroform, as digitalin is dissolved by it in all proportions, while digitin is insoluble.

The proportion of digitalin in *Digitalis* grown in different countries, has been made the subject of special investigation by Prof. S. P. Duffield, of Detroit. (94–1868.)

Digitalin is very bitter to the taste. It powerfully irritates the nostrils, and is an active poison. If digitalin be moistened with strong sulphuric acid and then exposed to the vapors of bromine, it assumes a purple color, which is darker or lighter according to the proportions employed. Hydrochloric acid produces with digitalin a very intense emerald green color.

One-fourth of a milligram is sufficient to produce the ordinary poisonous effects of digitalis. A milligram produces, in from three to five days, a marked change in the circulation. Three milligrams produce most dangerous effects within 24 hours.

EMETIA. 167

It is much to be desired that physicians substitute this crystalline substance, which is invariable, for the amorphous digitalin, which varies greatly, both as to character and effectiveness. Tardieu places digitalin among the hyposthenic poisons.

Poisoning by digitalin has often been produced through imprudence.

The *upas antiar*, with which the Indians poison their arrows, is obtained from the *Antiaris toxicaria*.

EMETIA.

This body is obtained from the roots of the ipecacuanha, *Cephælis ipecacuanha;* it also exists in the *Richardsonia braziliensis*, in the *Phsychtria emetica*, and in the roots of the *Cainca* (madder tribe). These materials, reduced to a powder, are treated with concentrated alcohol, and the alcohol then distilled off. The extract is diluted with five times its volume of water, and filtered. To the filtrate 2 per cent. of caustic potassa is added, and this mixture agitated with chloroform. The chloroform is decanted and distilled ; the emetia separates out. It is dissolved in dilute sulphuric acid, and precipitated from the solution with ammonia. A. Glenward (105-[3] 6—201) gives $C_{15}H_{22}NO_4$ as the formula of emetia.

It is amorphous, yellowish, fusible at 50°, soluble in water and alcohol. Its solutions are slightly bitter. It is a very weak base, and its salts are not crystalline. A few centigrams suffice to produce vomiting.

CANTHARIDIN

is a very poisonous crystalline substance, obtained from Spanish flies, (*Lytta vesicatoria*, and other varieties) and has the composition $C_5H_6O_2$. It is present in nearly all parts of the flies, varying in amount from 0.5 to 1.2 per cent. R. Wolff has of late given this substance a very full investigation. (95, May, '77–102.)

CAFFEINE (CAFFEIA) OR THEINE (THEIA).

$$C_8H_{10}N_4O_2, H_2O.$$

Alcohol is added to a mixture of 5 parts coffee and 1 part slacked lime, until nothing further is dissolved, and the solution distilled. The residue is treated with water, which causes an oil to separate out. The watery liquid furnishes crystals which are purified by treating with animal charcoal, and recrystallizing in hot water.

The extractive matters of the *kola-nut* and *maté* possess the same properties as caffeine.

Caffeine crystallizes in fine needles, fusible at 178°, and is volatile at a slightly higher temperature. These crystals are but little soluble in ether and cold water, yet dissolve very readily in alcohol and boiling water.

It is remarkable that the instinct of man should have led him to select, as the bases of common beverages, just the four or five plants, which out of many thousands are the only ones, as far as we know, containing caffeine.

It is recognized by boiling with fuming nitric acid; a yellow liquid is produced. On being evaporated to dryness, and ammonia added to the residue, a purple coloration is produced, resembling murexide. (p. 125.) *Amalic acid* and *Cholestrophan* are products of the action of oxidizing agents upon caffeine; bodies linking this alkaloid to the uric acid group.

THEOBROMINE.

There is extracted from the caco, *Theobroma cacao*, a principle crystallizing in microscopic crystals, volatile at 295°, soluble in alcohol and ether, and slightly so in water. It furnishes salts which are decomposed by water. It is called *theobromine*; its formula is $C_7H_8N_4O_2$.

PICROTOXIN.

$$C_5H_6O_2.$$

From the Indian berry, *Cocculus Indicus*, there is extracted a white crystalline matter of extreme bitterness, called *picrotoxin*, (from πικρός bitter τοξικόν.) This body is neutral, difficultly soluble in water, and easily soluble in alcohol and ether; its solutions are levogyrate.

The physiological action of picrotoxin is analogous to that of strychnia, but it differs from it in that it renders the action of the heart slower, and produces vomiting.

Prof. J. W. Langley, of Pittsburg, has contributed

much to (87–1862) our knowledge of the chemical character of picrotoxin.

POLYATOMIC ALKALOIDS.

There are polyatomic bases which are to the monatomic bases what polyatomic alcohols are to monatomic alcohols.

They are built upon the type of several molecules of ammonia, or condensed ammonia, in the same manner that polyatomic acids and alcohols are derived from several molecules of water.

Clocz obtained the former by the action of ethylene bromide upon potassa dissolved in alcohol.

Hoffmann established their true formula. They are called *polyamines*.

EXAMPLE.

$$\text{Ethylenic diamine. } N_2 \begin{cases} C_2H_4' \\ H_2 \\ H_2 \end{cases}$$

$$\text{Diethylenic \quad " } \quad N_2 \begin{cases} C_2H_4'' \\ C_2H_4'' \\ H_2 \end{cases}$$

$$\text{Triethylenic \quad " } \quad N_2 \begin{cases} C_2H_4'' \\ C_2H_4'' \\ C_2H_4'' \end{cases}$$

UREA.

$$CH_4N_2O = N_2 \begin{cases} CO'' \\ H_2 \\ H_2 \end{cases}$$

Rouelle, Jr., was the first to obtain this body in an impure state from urine.

Fourcroy and Vanquelin first obtained it pure.

Woehler, in 1828, prepared it artificially by a remarkable synthesis, the first attempt to form a body synthetically. Urea forms the chief constituent of the urine of mammalia, amounting to nearly one-half of the solid constituent; a small proportion of urea is found in all the fluids of the body.

It is an excretory product, as the hydrogen and carbon which have taken their part in the body, escape mainly in the form of water and carbon dioxide, so the nitrogen is eliminated from the system chiefly in the form of urea.

Urea may be extracted from urine by evaporating this liquid to one-tenth its volume and adding, after it has become cold, an excess of nitric acid. Brown crystals of nitrate of urea are formed: these are drained, expressed, re-dissolved in water and boiled with animal charcoal. This solution is filtered, and on evaporation it deposits crystals of nitrate of urea. This salt is then dissolved in as small a quantity of water as possible, and the solution treated first with barium carbonate, then with a strong solution of potassium carbonate; urea is set free and barium and potassium nitrates formed. The above mentioned salts are added as long as effervescence is produced; the liquid is then evaporated to dryness, and the residue treated with absolute alcohol, which dissolves only the urea. (J. E. Loughlin, 100-5-362.)

The synthetic method employed by Woehler, consists in preparing cyanate of ammonia, which body is isomeric with urea.

CYANATE OF AMMONIUM = $H_4CN_2O = NH_4-O-CN$.

This substance changes spontaneously into urea.

Heat, upon an earthen plate, 28 parts of potassium ferrocyanide and 14 parts of manganese dioxide, both finely pulverized, and dry until the mixture becomes pasty; when cold, the mass is pulverized and treated with water, and 20 parts of ammonium sulphide added to the liquid, which is now evaporated in a water bath, and the residue treated with boiling alcohol. On evaporating the alcoholic solution, crystals of urea are deposited. Urea is also obtained as a product of other reactions. It crystallizes in prisms of the tetragonal system; these crystals are colorless, without odor, and have a cooling taste.

It is soluble in its own weight of water at 15°, in an equal weight of boiling alcohol, and in 5 parts of cold 80 per cent. alcohol; it is difficultly soluble in ether. Its solutions are neutral.

Urea fuses at 120°; at about 150° it is decomposed, yielding ammonium carbonate, *ammelide*, $C_3OH_5N_5$, and *biuret*, $C_2O_2H_5N_3$.

Oxydizing agents decompose urea. Chlorine also decomposes solutions of urea in the following manner:

$$3Cl_2 + H_2O + CH_4N_2O = 6HCl + N_2 + CO_2.$$

Urea heated to 140° with water in sealed tubes, is transformed into ammonia and carbon dioxide:

$$H_2O + CH_4N_2O = CO_2 + 2NH_3.$$

This transformation likewise occurs when urea is heated with strong sulphuric acid, or fused with potassa, also, spontaneously, in presence of the nitrogenous matters of the urine.

Urea does not appear to unite with all acids. It has not yet been combined with carbonic, chloric, lactic or uric acids. The nitrate, chloride and oxalate of urea are crystalline.

Urea forms combinations with mercury, silver, and sodium oxides, also with mercuric and silver nitrates, etc.

NATURAL FATS AND OILS.

The fatty bodies are very widely distributed throughout the vegetable and animal kingdoms. Some are liquid, others are more or less solid. Certain oils remain liquid exposed to the air, as olive oil; others oxydize and thicken, as linseed oil, poppy oil, and nut oils; the latter are called *siccative* oils, and are used in the manufacture of varnishes, printers' ink, oil cloth, also in paints.

Fats and oils are insoluble in water; they are among the very few bodies which are wholly insoluble in this menstruum; they are also, in general, difficultly soluble in alcohol. They generally dissolve in ether, and the liquid hydro-carbons. Their specific gravity is less than that of water..

Heat destroys them; acrolein is usually formed associated with other products.

Since oil and water repel each other, many other substances may be protected from moisture by simply coating them with oil. Shoe-leather may be rendered water-proof and iron protected from rusting by greasing. Wood, saturated with oil, will last for a long time when buried in moist ground.

STEARIN OR STEARINE, (from $\sigma\tau\acute{\epsilon}\alpha\rho$, suet) $C_{57}H_{110}O_6$, is prepared by melting suet in turpentine; the two other proximate principles present, are precipitated,

FATS AND OILS. 175

while the stearine remains in solution. It is separated from the liquid by water, and purified by several recrystallizations in ether; it fuses at 71°, and solidifies at 50°.

Berthelot has reproduced stearine synthetically, by heating 3 parts of stearic acid with one part of glycerine, in a sealed tube.

This synthesis, as well as other researches, establishes the fact that the neutral fats are compound ethers of glyceryl, and the fatty acids.

On account of the heat generated by oxidizable oils when exposed to the air, frequent instances of spontaneous combustion occur when cotton rags, or waste soaked with oil, are allowed to remain in a heap.

Fats, especially if mixed with nitrogenous matter, become acid, *rancid*. The chemical nature of this change is not entirely understood.

OLEIN OR OLEINE, is the chief constituent of olive oil and fish oil. Berthelot has shown, by the action of oleic acid on glycerine, that natural oleine is a mixture of monoleine, dioleine, and trioleine. Oleine heated with a small quantity of mercury nitrate, or any other body capable of furnishing nitric oxide, becomes solid, owing to the transformation of the oleine into an isomeric body, *elaidine*. Siccative oils contain, instead of oleine, another principle called *elaine*.

Neutral fatty bodies and other ethers of glycerine are decomposed by alkaline solutions; a combination with water takes place, glycerine and fatty acids are formed. We may take as an example, stearin.

$$3KHO + C_{57}H_{110}O_6 = 3(KC_{18}H_{35}O_2) + C_3H_8O_3.$$

Alkalies, therefore, react upon the ethers of glycerine in the same manner as do the ethers of glycol and ordinary alcohol. This reaction is called *saponification*, and soaps are salts formed by stearic, margaric, and oleic acids, with a metal.

SOAPS. STEARINE CANDLES.

The only soluble soaps are those whose base is potassa or soda. Soda soaps, those ordinarily in use, are hard, while potassa soaps are soft. On adding to an aqueous solution of soap a solution of a metal, a precipitate is formed which is the soap of the metal employed ; thus the precipitate which common water produces in soap is a lime soap.

Ordinary soap is made by boiling fats of inferior quality with an alkaline solution. When the oil is completely decomposed the soap is precipitated by salt water, in which soap is insoluble.

Stearine candles have hitherto been made by saponifying suet or tallow with lime in the presence of boiling water. At present the amount of lime employed in the saponification is considerably diminished (amounting to only 4 per cent.) by operating at a temperature of 150°.

The saponification of fats of inferior quality is also effected by means of sulphuric acid instead of lime; this acid forms with the fatty acids, double or conjugate acids, which are decomposed by water. The decomposition of fats into their constituents, the fatty acids and glycerine, for the manufacture of candles, is at present effected on a large scale by simply heating the fats with steam under pressure, and at a temperature of 260°. This is the celebrated process of the American inventor, Tilghman, to whom the wonderful "sand blast" is also due.

This decomposition of fats is most remarkable, as, by the same process, only at a lower temperature, Berthelot obtained a result exactly the reverse, causing stearic acid and glycerine to reform stearine by simple direct synthesis.

STEARIC ACID, $C_{18}H_{36}O_2$, is crystalline, insoluble in water, soluble in alcohol and ether, and melts at 70°. It unites with the bases; its alkaline salts alone are soluble.

MARGARIC ACID, having the formula $C_{17}H_{34}O_2$, (from μαργαρον, a pearl, owing to its pearly lustre) is crystalline. It melts at 60° and forms salts with the metals.

OLEIC ACID, $C_{18}H_{34}O_2$, is an oil becoming colored in the air and converted into an acid called *elaidic acid*, which is fusible at 44°, in contact with a small quantity of hyponitric acid.

These three acids, stearic, margaric, and oleic, are those that, with glycerine, constitute most of the natural fats, or glyceryl ethers.

LEAD PLASTER is essentially a lead-soap compound of plumbic oleate.

CROTON OIL.

This oil is extracted from the seed of the *Croton tiglium* of the family of euphorbiaceæ.

The seeds are ground and expressed, or they are treated with ether, which is afterwards driven off by distillation.

This oil is yellowish, very bitter, and possesses a disagreeable odor. Alcohol and ether dissolve it. It produces blisters whenever it comes in contact with the skin, and is a drastic poison.

Pelletier and Caventou have extracted from this oil an acid body, $C_4H_6O_2$, denominated *crotonic acid*.

COD-LIVER OIL.

This oil is extracted from the liver of the cod, and several other species of the genus *Gadus*. Two processes are employed for its extraction; either the oil is obtained by putrefaction, in which case the oil separates out naturally, or the livers are cut into small pieces and heated in large pans, then placed in cloth sacks and pressed. It is of a brownish color. A white oil is sometimes sold, which has been bleached by treatment with weak lye and animal charcoal. The efficiency of this latter oil is much less than that of the natural oil.

There has been found in this oil 3 to 4 thousandths of iodine, and a small quantity of phosphorous; and its medical qualities are thought to be due to these

WAX. 179

two substances, but it is probable that its efficiency is more frequently due simply to its fatty character.

BUTTER.

Ordinary Butter. Butter contains stearic, margaric, oleic, and butyric acids, and several other proximate neutral principles. Its density is 0.82. It dissolves in 30 per cent. of boiling common alcohol. The odor which it emits on becoming rancid is due to the liberation of fatty acids.

"*Oleo-margarine*" is artificial butter, consisting mainly of oleine and margarine obtained from suet or lard.

SPERMACETI.

This substance which is formed in peculiar cavities in the head of the sperm whale, and is a neutral fatty body sometimes employed in pharmacy. It is an ether, which, on saponification, produces a fatty acid called *ethalic* acid, and a monatomic alcohol, *ethal*.

$$H_2O + \underbrace{C_{32}H_{64}O_2}_{\text{Spermaceti.}} = \underbrace{C_{16}H_{31}OHO}_{\text{Ethalic Acid.}} + \underbrace{C_{16}H_{34}O}_{\text{Ethal.}}$$

WAX.

Yellow bees-wax is obtained by submitting honey-comb to pressure, then fusing the same under boiling water. It is bleached by being cut into thin cakes and exposed to the air and sunlight. Thus prepared it fuses at 62°. Mixed with 3 per cent. of oil of sweet almonds it forms a *cerate*, used in pharmacy.

On being treated with alcohol it separates into two proximate principles: one, soluble in this liquid, is acid, and is called *cerotic acid*, having the formula $C_{27}H_{54}O$; the other, which is but slightly soluble, is called myricin. The latter is a compound ether, and is decomposed by bases into an acid, *ethalic acid*, and an alcohol, *melissic alcohol*, $C_{30}H_{62}O$.

CASTOR OIL.

This oil is extracted from the *Ricinus communis*, a plant of the family of Euphorbiaceæ.

The castor-oil beans are hulled, pulverized, and the pasty mass obtained subjected to strong pressure. This oil is slightly yellow. Its density is 0.926 at 12°, and it remains liquid at a temperature of −18°. It is very soluble in alcohol, a characteristic which distinguishes it from most other oils.

This oil is also an ether of glycerine; the acid which it contains is *ricinoleic acid*, $C_{18}H_{34}O_3$.

SUGARS.

The general name of *sugars*, by some regarded as polyatomic alcohols, is given to bodies which are capable of fermenting, that is, of decomposing directly or indirectly into different products, of which the principal ones are alcohol and carbon dioxide. Fermentation requires the presence of certain microscopic plants, and, according to Pasteur, is a phenomenon correlative with the vital development of these organisms. This, however, has been latterly disproved by Tyndall.

Sugars may be divided into three classes. In the first are those in which the proportion of hydrogen is more than sufficient to convert the whole of the oxygen into water. It contains:

Mannite, $C_6H_{14}O_6$, extracted from manna.

Dulcite or *mélampyrite*, $C_6H_{14}O_6$, found in Madagascar.

Pinite, $C_6H_{12}O_5$, extracted from a Californian pine tree.

Quercite, $C_6H_{12}O_5$, extracted from acorns.

These bodies do not ferment with beer yeast alone; but in presence of certain ferments and calcium carbonate they furnish alcohol, carbon dioxide, and hydrogen.

Sugars of the second and third class contain hydrogen and oxygen in the proportions to form water.

The second class includes the glucoses, *isomeric bodies*, whose general formula is, $C_6H_{12}O_6$. Among these are:

Ordinary *Glucose* or *grape sugar*.

Levulose, associated with glucose in the form of inverted sugar.

Maltose, obtained from malt.

Galactose, obtained by treating sugar of milk, or gums, with dilute acids.

Eucalin, obtained by the action of maltose on beer yeast.

Sorbin exists in the berries of the mountain ash.

Inosite is found in the embryo of young plants and in the fluids of flesh.

Lactose or Sugar of Milk. The glucoses may be divided into two series. The first includes those bodies (ordinary glucose, levulose) which, on being oxydized, form saccharic acid, and on being hydrogenized by means of sodium amalgam, produce *mannite*. The second includes those substances (galactose, lactose) which, on oxydation produce mucic acid, and on hydrogenation furnish *dulcite*. The third class of sugars contains bodies whose general formula is $C_{12}H_{22}O_{11}$, and are called *saccharoses*, by Berthelot. It contains, besides cane sugar, three bodies called:

Melitose, an exudation of certain eucalypti.

Trehalose or *mycose*, extracted from the Turkish manna and certain mushrooms.

Melezitose, obtained from an exudation of the larch.

The sugars of the first two classes are placed by Berthelot among the polyatomic alcohols.

MANNITE.

$C_6H_{14}O_6$.

This body exists naturally in an exudation of various species of ash (*Fraxinus rotundifolia*), called manna, of which it forms the greater portion. It is also found in mushrooms, algæ, the sap of most fruit trees, onions, asparagus, celery, etc. It may be prepared by dissolving manna in one-half its weight of water, to which a small quantity of egg albumen is added, and the mixture brought to boiling and filtered. On cooling, colored crystals are deposited which are expressed and redissolved in hot water. This solution is mixed with animal charcoal, boiled and filtered while hot. The liquid deposits crystals on cooling. Mannite crystallizes in rhombic prisms and has a sweet taste. It dissolves in seven times its own weight of cold water, is slightly soluble in alcohol, and insoluble in ether. Its solutions are optically inactive.

Mannite fuses at about 165°; at about 200° it yields a certain quantity of a substance called *Mannitane*, $C_6H_{12}O_5$. It oxydizes in presence of platinum black, furnishing a non-crystallizable acid called *mannitic acid*. Boiling nitric acid converts it into saccharic and oxalic acids.

Mannite, treated with a small quantity of nitric acid, is changed into a body insoluble in water, called *nitro-mannite*, $\begin{Bmatrix} (C_6H_8) \\ (NO_2)_6 \end{Bmatrix} O_6$, which may be regarded as a compound ether.

Dulcite.—Dulcite is very analogous to mannite, but differs from it, in that it furnishes, with nitric acid, mucic acid.

GLUCOSES.

$$C_6H_{12}O_6.$$

These compounds may be considered as representative carbohydrates. Ordinary glucose (from γλυκυς, sweet,) or grape-sugar, is a crystalline substance, and is found in honey, figs, and various other fruits, together with another insoluble glucose. It has been found in small quantity in the liver and in most of the fluids of the body. It is obtained by the decomposition of salicine, tannin, and other substances, which, for this reason, have been named *glucosides*.

Vegetable cellulose, the envelope of many invertebrates (chitin and tunicin) and the glycogenous principle of the liver furnish glucose on treatment with dilute acids.

It is manufactured on a large scale by the action of starch upon dilute sulphuric acid. Water containing four to eight per cent. of sulphuric acid is placed in vats and heated to boiling by means of superheated steam. Before the water boils, starch mixed with water is added, and ebullition maintained as long as a small quantity of the mixture gives a blue reaction with iodine. The sulphuric acid is not changed during this transformation.

It is then saturated with chalk and the liquid allowed to become clear. It is decolored by passing through filters containing animal charcoal and evaporated to a density of 41° Baumé. The glucose crystallizes in compact masses. Often the liquid is evaporated to only 3° B., when a syrup is obtained known as *starch syrup*. Honey treated with cold concentrated alcohol, also furnishes glucose. The crystals of glucose are small, opaque, and ill defined.

They are represented by the formula $C_6H_{12}O_6,2H_2O$, but they may be obtained having the composition $C_6H_{12}O_6$ by precipitating the glucose in boiling concentrated alcohol. The water may also be driven off by heating the glucose to about 100°.

Glucose is soluble in a little more than its own weight of water. Weak alcohol dissolves it readily. It is slightly soluble in cold concentrated alcohol.

Its solutions turn the plane of polarization to the right. This rotatory power is feeble in the cold.

Glucose, heated to about 170°, acts in the same manner as mannite. Gelis has demonstrated that it loses a molecule of water; the body formed $C_6H_{10}O_5$, is called *glucosane*, $C_6H_{12}O_6 = C_6H_{10}O_5 + H_2O$. It reproduces glucose on being boiled with acidulated water. If glucose is boiled with dilute nitric acid, saccharic and oxalic acids are formed. Fuming nitric acid forms with glucose a very explosive compound.

Hydrochloric acid turns it brown. With dilute sulphuric acid it furnishes a double acid (*sulphoglucic acid*); with strong sulphuric acid, carbon. Glucose oxydized with care, furnishes saccharic acid.

Heated to 100° with butyric, or various other acids,

it loses water, and the glucosane formed reacts upon the acid, forming an ether, *saccharide*, or *dibutyric glucosane*,

$$\left.\begin{array}{l}(C_6H_6)\\(C_4H_7O)H_2\end{array}\right\} O_5.$$

This body, as well as other *saccharides*, are decomposed under the action of boiling acidulated water, into an acid and glucose.

Glucose combines, with sodium chloride, forming several crystalline compounds; it also forms unstable compounds with the metallic bases,

$$CaC_6H_{10}O_6$$
$$BaC_6H_{11}O_6, \text{etc.}$$

Péligot has shown that the solutions of these glucosates are gradually changed into salts of a special acid called *glucic acid*, whose formula is

$$C_{12}H_{18}O_9.$$

Cupric acetate boiled with glucose is reduced to the state of suboxide.

This action, which is very slow with salts of copper with inorganic acids, becomes rapid and complete in presence of alkalies. On adding glucose to a solution of copper sulphate, this salt is not precipitated by potassa. If, however, the liquid is heated, it deposits cuprous oxide. (Trommer's test.) This reaction is more delicate with copper salts, whose acids are organic. A mixture is used of copper sulphate, Rochelle salt and soda (Fehling), or a solution of copper tartrate in potassic hydrate. (Barreswil.)

Prof. W. S. Haines has found in glycerine a very desirable substitute for the tartrate in Fehling's test. The proportions employed by him for qualitative examinations are: cupric sulphate, 30 grains; potassic hydrate, 1½ drachms; pure glycerine, 2 fluid drachms; distilled water, 6 ounces.

LEVULOSE, $C_6H_{12}O_6$.

This name is given to a variety of glucose, which is found in many fruits. It may be obtained by boiling inulin with water, or, better, it can be prepared from cane sugar by the action of dilute acids. It differs from the other sugars in that its rotary power diminishes on heating.

GALACTOSE,

$$C_6H_{12}O_6.$$

This body is produced by boiling, for two or three hours, sugar of milk with water acidulated with sulphuric acid. It is soluble in water and insoluble in alcohol; nitric acid transforms it into *mucic acid*.

INOSIN, INOSITE OR MUSCLE SUGAR.

$$C_6H_{12}O_6 + 2H_2O.$$

This substance is found in many animal organs, and

is the chief constituent of the liquid which impregnates the muscles.

It may be prepared by first extracting the creatin from the muscles, then separating the inosic acid with baryta. To the liquid is then added a quantity of sulphuric acid sufficient to precipitate the whole of the baryta and the liquid treated with ether, which dissolves the foreign substances.

The aqueous solution is removed and alcohol added to it until a precipitate is formed. Crystals of potassium sulphate first separate out, then beautiful crystals of inosite. This substance has a sweet taste. At a temperature of 100° it loses two molecules of water. It dissolves in one-sixth of its weight of water while it is insoluble in ether and strong alcohol.

Inosite is without action upon polarized light. It is not converted into glucose by the action of dilute acids, and does not reduce copper salts. Mixed with milk and chalk it undergoes lactic fermentation. (Page 122.)

SACCHAROSES.

Ordinary Sugar,

$$C_{12}H_{22}O_{11}.$$

This body exists in a large number of plants, though it is almost exclusively extracted from the sugar-cane and beet-root.

The sugar-cane, *Arundo saccharifera*, contains 17 to 20 per cent. of sugar. To extract, the juice of the cane is first obtained by expressing. This juice represents 60 to 65 per cent. of the total weight of the cane, and would alter rapidly in the air if care were not taken to bring it rapidly to a temperature of 70°, and adding a quantity of lime. The juice soon becomes covered with foam and deposits different albuminoid and other matters, which are precipitated by the lime. It is decanted into pans and rapidly evaporated. The sugar crystallizes out, and the mother liquor is evaporated as long as it furnishes crystals. The thick liquid which remains is *molasses*. The sugar thus obtained is brown sugar, and is subsequently refined.

The beet-root most rich in sugar is that of Silesia. It contains about 10 per cent. of sugar. Sugar crystallizes in clinorhombic prisms. They may be readily obtained by slowly evaporating a solution of sugar.

The crystals of ordinary sugar are very small, as the syrup is made to crystallize quite rapidly. Cold water dissolves three times its weight of sugar; hot water dissolves it in all proportions, forming a syrupy liquid. It is not dissolved by cold alcohol or ether. Dilute alcohol dissolves it in proportion as it is more or less aqueous. Its solutions are dextrogyrate. Sugar melts at about 180°, and yields a liquid which solidifies to a vitreous, amorphous mass, called *barley sugar*, which becomes opaque and crystalline after some time.

If sugar is heated a little above this point, it is transformed into glucose and levulosane.

$$C_{12}H_{22}O_{11} = C_6H_{12}O_6 + \underbrace{C_6H_{10}O_5}_{\text{Levulosane.}}$$

At about 190° sugar loses water, becomes brown, and finally furnishes a substance which is commonly known as *caramel*. According to Gelis three products of dehydration are formed, *caramelane*, *caramelene* and *carameline*. At a temperature of 230° to 250° sugar is decomposed into carbon monoxide, carbon dioxide, carbohydrides and different empyreumatic products. Sugar is transformed slowly in the cold, and rapidly at 80°, in contact with dilute acids into *inverted sugar*, which is thus called on account of its inverted action upon polarized light. On prolonged ebullition the solution is rendered brown and ulmic products are formed. Sugar reacts with baryta water and lime water, forming different compounds called *sucrates* or *saccharates*.

SUGAR OF MILK.

The solutions of these sucrates are decomposed by carbon dioxide: sugar is reformed. Rousseau makes use of this fact in the manufacture of sugar on a very large scale.

Sugar does not ferment immediately in contact with beer yeast.

SUGAR OF MILK, LACTIN OR LACTOSE.

$$C_{12}H_{22}O_{11} + H_2O.$$

It is obtained from milk, by precipitating the casein with a few drops of dilute sulphuric acid, filtering and evaporating the liquid.

Crystals are deposited, which are purified by re-dissolving and treating with animal charcoal.

In Switzerland large quantities of sugar of milk are made by evaporating the whey which remains after the separation of the cheese.

The crystals of this body are rhombic prisms. This sugar is insoluble in ether and alcohol, and requires 2 parts of boiling and 6 parts of cold water for its solution.

Its solutions are dextrogyrate. At a temperature of about 140° it loses H_2O, and becomes brown at 160° to 180°.

In presence of sour milk and chalk it undergoes lactic fermentation.

Sugar of milk is extensively used in homœopathic pharmacy; also in the pepsin of commerce, and in saccharated extracts.

Reichardt has obtained from gum arabic a sugar distinct from ordinary sugar, a body though having the same formula. He names it *para-arabin*.

HONEY.

Honey is produced by the domestic bee (Apis mellifica), an insect of the order Hymenoptera.

It is separated from the wax by exposing the honeycomb to the sun, on wire nets; very pure honey is thus obtained.

The mass which remains is expressed, and this product is a second quality of honey, more colored and of a less agreeable taste and odor than the first. The comb is then heated with water to remove the remainder of the honey. The wax thus isolated is melted and run into moulds. Honey owes its sweet taste to several sugars. There is found in it a dextroyrgate, crystallizable glucose, and on removing this sugar there remains a viscid uncrystallizable liquid, which contains levulose. In addition to these, small quantities of ordinary sugar have also been found in honey.

GLUCOSIDES.

This name is given to certain bodies which have the property of forming various products by combining with water, among which is glucose, or some other saccharine matter.

This change is produced by the action of acids, bases, or by the action of ferments. We cite the following, but shall only study the most important:

Salicin, $C_{13}H_{18}O_7$, extracted from the bark of the Willow.

Amygdalin, $C_{20}H_{27}NO_{11}$, extracted from the Bitter Almond, *Amygdalus communis*.

Orcin, $C_7H_8O_2$, extracted from various Lichens.

Tannin, $C_{27}H_{22}O_{17}$, extracted from the Oak.

Phlorizin, $C_{21}H_{24}O_{10}$, extracted from the Apple, Pear, or Cherry tree.

Populin, $C_{20}H_{22}O_8$, extracted from Aspen leaves.

Arbutin, $C_{13}H_{16}O_7$, extracted from the leaves of the Uva-Ursa.

Convolvulin, $C_{31}H_{50}O_{16}$, extracted from the *Convolvulus orizabensis* and *schiedeanus*.

Jalappin, $C_{34}H_{56}O_{16}$, extracted from *Convolvulus orizabensis* and *scammonia*.

Saponin, a white amorphous powder whose solution is very frothy and of which the powder is very sternutatory.

Daphnin, $C_{30}H_{32}O_{18}$, the crystalline matter extracted from the bark of the Ash (*Fraxinus excelsior*).

Cyclamin $C_{20}H_{21}O_{10}$, extracted from the tubercles of the *Cyclamen europæum*.

Quinovin, $C_{30}H_{48}O_8$, a resinous, bitter matter, soluble in alcohol, existing in the bark of the *Quina nova* and other cinchonas.

Solanin, $C_{43}H_{71}NO_{16}$. This has already been studied, (page 165).

Esculin, $C_{30}H_{34}O_{19}$, extracted from the bark of the Horse Chestnut.

Quercitrin, $C_{29}H_{30}O_{17}$, from the bark of the yellow oak (*Quercus tinctoria*).

Coniferin, $C_{16}H_{22}O_8$, from the *Larix europaea*, etc.

Vanillin, from the Vanilla bean, and recently obtained artificially (60-74-608).

SALICIN, $C_{13}H_{18}O_7 + H_2O$.

This body crystallizes in white needles, fusible at 120°, insoluble in ether, soluble in alcohol and water. These solutions are levogyrate and very bitter. It is used as a febrifuge, but is of little value in well defined intermittent fevers.

It has as a distinguishing chemical character, the property of becoming red with sulphuric acid.

Under the action of dilute sulphuric, or hydrochloric acid, or even with emulsin, salicin is decomposed. With the latter the reaction is:

$$C_{13}H_{18}O_7 + H_2O = \underbrace{C_6H_{12}O_6}_{\text{Glucose.}} + \underbrace{C_7H_8O_2}_{\text{Saligenin.}}$$

In contact with cold nitric acid it loses hydrogen, and a body is formed called *helicin*, $C_{13}H_{16}O_7$.

When treated with oxydizing agents, it gives off an odor which is identical with that of the essence of meadow sweet (*Spirea ulmaria*).

This body is produced especially when salicin is treated with a mixture of sulphuric acid and potassium bichromate, and is also known by the name of *hydride of salicyl*.

Its formula is identical with that of benzoic acid, $C_7H_6O_2$, but it has not the properties of this acid.

It is an aromatic liquid, boiling at 196°, and has the property of oxydizing spontaneously, giving rise to an acid called *salicylic acid*, $C_7H_6O_3$.

Salicin, treated with fused potassa, furnishes potassium oxalate and salicylate. Cahours has shown that essence of *Gaultheria procumbens*, a heath of New Jersey, contains, besides, an isomer of the essence of turpentine, a sweet-scented liquid, boiling at 220°, which is salicylic methyl ether, and is re-converted, in contact with alkalies, into methyl alcohol and salicylic acid : it may be produced artificially by treating wood spirit with a mixture of salicylic and sulphuric acids.

Salicylic or oxybenzoic acid has been lately produced by Kolbe (56 –'74 –22), by a remarkable synthesis in acting on carbolate of sodium with CO_2.

$$2C_6H_5ONa + CO_2 = C_6H_6O + C_7H_4O_3Na_2.$$

Sodium phenol.   Sodium salicylate of sodium.

It has now come to be a very important article in pharmacy and in the arts, on account of its efficiency as an antiseptic, equaling or surpassing carbolic acid (phenol), yet without the unpleasant odor of the latter body, or its toxical qualities. As of considerable importance theoretically, it should be stated that Herrmann has very lately (60–April, '77) obtained salicylic acid by the action of sodium upon succinic ether.

TANNINS.

This is the name given to different principles existing in plants, which are characterized by the following properties:

1st. They give, with ferric salts, a black coloration approaching blue or green.

2d. They precipitate solutions of albuminoid substances, particularly those of gelatine.

The principal ones are:

Tannin of oak, $C_{27}H_{22}O_{17}$.
" " cachou (catechin or catechic acid).
" " quinquinia (quinotannic acid).
" " coffee (caffetannic acid).
" " fustic (morintannic acid).

Oak tannin is best prepared from gall-nuts which contain much more than does the bark. The nuts are pulverized and submitted to the action of commercial sulphuric ether, which is made aqueous. This ether may be replaced with advantage by a mixture of 600 grams of pure ether, 30 grams of 90 per cent. alcohol, and 10 grams of distilled water for every 100 grams of gall-nuts. After twenty-four hours the apparatus contains two layers of liquid; the upper one is ether, containing but little tannin, while the lower one is a very strong aqueous solution of tannin.

The lower layer is removed and evaporated in an

oven on shallow plates. There remains an amorphous spongy substance, very soluble in water, less soluble in alcohol, and almost insoluble in ether. This residue is very astringent and slightly acid.

Solutions of tannin give a white precipitate with tartar emetic.

It precipitates solutions of the alkaloids, and coagulates blood.

With solutions of gelatin it gives a voluminous precipitate, soluble on heating in an excess of gelatin.

Tannin forms, with fresh hide, an imputrescible compound, which is *leather*. The art of tanning is based on the action of oak-bark tannin on hides from which the hair has been removed, usually by lime.

GALLIC ACID. In solution, tannin is gradually decomposed, the liquid becoming covered with mould.

Carbon dioxide is disengaged and an acid, called *gallic acid*, is formed.

This transformation does not take place if all air is excluded; and the air *alone* is not sufficient. It requires the presence of a mycelium of a mucedin conveyed to the liquid either by the air or in some other manner.

This transformation is, like alcoholic fermentation, a phenomenon correlative with the development and growth of an organism. On boiling tannin with water acidulated with hydrochloric or sulphuric acid, it is decomposed into glucose and gallic acid:

$$C_{27}H_{22}O_{17} + 4H_2O = 3(\underbrace{C_7H_6O_5}_{\text{Gallic acid.}}) + \underbrace{C_6H_{12}O_6}_{\text{Glucose.}}$$

Gallic acid is deposited as the liquid becomes cool. It is purified by redissolving and treating with animal charcoal, and recrystallizing.

Gallic acid, $C_7H_6O_5 = \left. \begin{array}{c} C_7H_2O \\ H, H_3 \end{array} \right\} O_4$, crystallizes in silky needles, soluble in three parts of boiling water, but little soluble in cold water. This solution, on standing in the air, becomes altered after a long time, carbon dioxide is disengaged and the solution turns brown; alkalies accelerate this change.

Gallic acid produces a blue color with ferric salts, and precipitates tartar emetic, but does not precipitate gelatin when pure, nor the alkaloids.

Mixed with pumice-stone and heated to 210° it produces a beautiful sublimate of *pyrogallic acid*, carbon dioxide being liberated at the same time.

$$C_7H_6O_5 = C_6H_6O_3 + CO_2.$$

This body occurs in colorless, acicular crystals, fusible at about 115°, and soluble in 2.5 parts of water. Its solution absorbs oxygen from the air, in presence of alkalies, and becomes quite brown.

It reduces gold and silver salts, and forms unstable compounds with certain acids. It may properly be placed among the phenols. This body is employed in photography, and in the laboratory. Mercadante (47–'74–484) finds that gallic acid is injurious to vegetation, inasmuch as it combines with the mineral food of the plant rendering it insoluble.

Grimaux was the first to consider gallic acid as tetratomic and monobasic (77 620).

VEGETABLE CHEMISTRY.

At the moment when the radicle of a plant appears above the ground, its vital phenomena undergo a marked change.

The plant decomposes carbon dioxide, water and certain nitrogenous compounds furnished by the soil, and grows by retaining carbon, hydrogen, nitrogen and a little oxygen, and returns to the air the greater part of the oxygen derived from the carbon dioxide, water and nitrogenous compounds.

Bonnet observed, in the last century, that leaves, exposed to the sun in areated water, disengage a gas, which Priestly showed is oxygen. Sennebier discovered that this oxygen is derived from carbon dioxide. De Saussure verified these facts, and demonstrated that this decomposition of carbon dioxide does not take place in the dark, and that the green portions of the plant alone are capable of effecting the change.

J. Belluci (9-78-362) has lately shown that, contrary to former belief, none of the oxygen exhaled by plants is in the form of ozone.

EXPERIMENT.—Place a few leaves in a flask half full of water containing carbon dioxide, "soda water," invert the flask over a glass of water, and expose it to the sunlight, after having covered it, if the sun is very hot, with a sheet of transparent paper; minute bubbles will soon be seen to form on the leaves, as small as the point of a pin, will increase in size, unite and mount to the upper part of the flask. Transfer this gas to a test-tube, and, on examination, it will be found to be oxygen. Substitute for this flask an opaque vessel, or perform the experiment in the dark, and the carbon dioxide will not be altered in the least.

Where do the plants find this carbon dioxide? Chiefly in the air. Boussingault, in order to demonstrate this, placed under a bell-glass some peas planted in calcined sand; he watered them with pure distilled water, and passed air into the glass; the peas grew, flowered and bore fruit.

Now the substance of these peas contained carbon hydrogen and nitrogen, in much greater quantity than the seed from which they grew, consequently these constituents were taken from the air and water.

If, however, the air be made to pass through an alkaline solution before escaping from the vessel, no carbon dioxide is absorbed, which also proves that the carbon dioxide existing in the air has been removed by the plant. The plant takes up, in the same manner, carbon dioxide from the water which passes from the soil into its roots.

Plants are also capable of decomposing water, in fact. Collin and W. Edwards have proved that the submerged stems of the *Polygonum tinctorium* and certain mushrooms, exhale hydrogen.

On the other hand, Payen has proved that the hydrogen exceeds the oxygen in the woody parts of plants, and, indeed, many substances produced by plants, as oils and resins, are very rich in hydrogen. In short, the oxygen contained in the plant would not be sufficient to oxydize or transform into water the whole of the hydrogen it contains, consequently it must be admitted that water is decomposed by plants. The conditions under which this change takes place have not as yet been determined.

The experiment of Boussingault proves, as Ingenhousz has claimed, that the air furnishes the plant with nitrogen; but where does this nitrogen come from? Is it taken by the plant from the free nitrogen of the atmosphere? or is it derived from the nitric or nitrous acids, or from the ammonia contained in the atmosphere, or, in one word, from the nitrogenous compounds existing in the air?

Boussingault has shown that while certain families of plants, principally the common vegetables, derive from the air a large quantity of nitrogen, even taking up free nitrogen, others, the cereals for instance, derive nitrogen chiefly from the soil; for, on causing clover and wheat to grow in calcined sand in presence of air deprived of its nitrogenous compounds, and distilled water, he observed that the clover took up carbon, hydrogen, water and nitrogen, while it appears that the wheat obtained from the air carbon and water only.

Nitrogen, which is present in the air in the form of ammonium nitrate, is absorbed by all plants. Direct experiments have shown that the salts of ammonium, especially ammonium nitrate, constitute an excellent compost, and consequently this nitrate can lose its oxygen, or become reduced in the plant.

Now, it is known that urea and animal excreta are transformed into ammoniacal compounds on exposure to the air; therefore, in order to obtain a good crop, even with plants which take up the nitrogen of the air, it is necessary to employ manures which furnish not only easily assimilated nitrogen, but those which, besides, furnish the plant with soluble organic compounds and the mineral substances necessary for its development and growth. Of these latter there is required for the plant, potassium and calcium chlorides, sulphates, phosphates, etc.

With the four elements, carbon, hydrogen, nitrogen, and oxygen, nature forms an infinite variety of compounds by mysterious methods, to which we have not, as yet, the key, but of which synthetical research gives us some idea. Thus, with carbon dioxide and water, Berthelot produces formic acid; with formic acid he obtains alcohol, and subsequently acetic acid. Pasteur also has shown that glycerine, one of the principles of fat, is produced in the process of fermentation and that a complex acid, succinic acid, is also formed under the same circumstances. However, we are far from knowing how to produce those substances which nature forms at ordinary temperatures, and *with only four elements*. What wondrous chemistry is that of the plant, fitted by an all-wise Creator to elaborate with such simple materials, the beauteous violet, the fragrant rose, or the luscious fruit!

VEGETABLE CHEMISTRY. 203

By combining six atoms of carbon with five atoms of water, nature forms either the woody principle, *cellulose*, or the essential constituent of the potato, *starch*. By uniting ten atoms of carbon with sixteen atoms of hydrogen, she produces, in the orange and in the pine, two essences or oils very different in character. By associating the four organic elements she forms the most different substances, the nourishing cereal as well as the most deadly strychnia; and often products as unlike as these are found side by side in the same plant.

Thus the plant is a structure which decomposes carbon dioxide, water, and compounds of nitrogen; which forms its substance out of carbon, hydrogen, nitrogen, and a part of the oxygen of these compounds, and which exhales oxygen. Hence, chemically, it would be proper to call the plant *a reducing apparatus*.

We should add that the flowers and portions of plants not green, also the buds in developing, produce an exhalation of carbon dioxide, and that during germination, and especially during the time of flowering, a sensible amount of heat is disengaged. As a result of this elevation of temperature, there is produced in plants some slight oxydation or combustion, as in the respiration of animals.

Hence, we must conclude that plants and animals, in many circumstances at least, deport themselves in a similar manner.

Many experimenters, and especially Dutrochet and Garreau, go further, and say that plants and animals respire in an identical manner, and according to their theories all living creatures take up oxygen and exhale carbon dioxide.

The experiments of Garreau especially deserve attention. He placed branches, detached or affixed to the plant, in vessels full of air, and exposed them to a diffused light. The volume of the air was known and the oxygen absorbed was determined by a special contrivance; the carbon dioxide produced was removed by placing in the vessel an alkaline solution of known weight. Thus the variations of these gases were carefully studied.

As a result of his experiments Garreau claimed to have established that both in the dark and in the light, there is an absorption of oxygen and an exhalation of carbon dioxide, but the amount of carbon dioxide collected does not represent the amount really exhaled, as the greater part is reduced at the moment of liberation. From these facts it would appear that in all living creatures the same phenomenon of respiration takes place, which consists in a consumption of oxygen and an exhalation of carbon dioxide.

This phenomenon is associated with another; viz., assimilation or nutrition. It is here that the difference, indeed a complete opposition, between the two kingdoms is established. The plant grows by reducing, under the influence of heat and sunlight, carbon dioxide, water and nitric acid, by accumulating carbon, hydrogen, nitrogen and by exhaling the greater part of the oxygen. The animal, on the other hand, forms its substance from that of the plant, oxydizing, or consuming, the vegetable products with the oxygen of the air exhaled by the plants; it reduces the complex products formed in the vegetable to the state of carbon dioxide, water and ammonia; thus the animals supply the plants with food, receiving in turn nourishment from them. Those desirous of further studying this and other interesting topics relating to Vegetable Chemistry, will find very valuable the works of Prof. S. W. Johnson, " How Crops Grow," and " How Crops Feed"; also Prof. John C. Draper's article in Am. Jour. Sci. and Arts, Nov. 1872, entitled "Growth of Seedling Plants."

ORGANIZED SUBSTANCES.

Among the chemical substances of which we have spoken certain ones participate more in vital phenomena, and have more definite physical structure than do others.

These are designated as *organized or organizable substances*, the term *organic* being reserved for the definite compounds studied in organic chemistry. All these substances play an important part in the vegetable kingdom, forming the network of vegetable tissue, as cellulose or as starch, etc.

CELLULOSE OR CELLULIN, $(C_6H_{10}O_5)_n$.

On examining a young plant under the microscope,

we observe that it is built up of little cells and minute, diaphanous ducts or vessels filled with sap and air. The material of which these tissues are composed is called *cellulose*. The pith of the elder, cotton fibre, and paper are almost exclusively composed of this substance.

Cellulose is a carbo-hydrate; $C_6H_{10}O_5$, is the formula, ordinarily given to it, although a multiple formula at least three times as large, or $C_{18}H_{30}O_{15}$ is necessary to explain certain reactions with nitric acid.

EXPERIMENT. Pure cellulose may be obtained in the following manner: cotton, linen or paper is treated with dilute alkaline solutions, washed and immersed in weak chlorine water; finally it is submitted to the action of various solvents, as water, alcohol, ether and acetic acid until nothing more is dissolved.

This substance is solid, white and insoluble. It is destroyed at a red heat, producing carbon and numerous carbohydrides, gaseous and liquid, which distil over. With monohydrated sulphuric acid it produces a colorless, viscid liquid, which contains, at first, an insoluble substance having the properties of starch and yielding a blue color with iodine. If the action of the acid is continued, the whole is dissolved and the same products are obtained as in the case of starch when brought in contact with sulphuric acid, i. e. dextrin and glucose. To separate the latter substance, it is simply necessary to saturate the acid with chalk and evaporate the liquid.

Concentrated hydrochloric acid produces the same

CELLULOSE. 207

effect. If paper be immersed for *an instant* only in sulphuric acid, diluted with half its volume of water, and carefully washed, it acquires the toughness of parchment. Paper thus prepared is frequently employed in experiments on dialysis; it is also much used by pharmacists to cover the stoppers of bottles. It is known in commerce as *vegetable parchment*.

GUN COTTON OR PYROXYLIN.

Gun cotton was first made by Schoenbein, in 1846.

To prepare it cotton is plunged for two or three minutes into fuming nitric acid, or, better, into a mixture of 1 vol. nitric acid (of a density of 1.5), and 2 vols. of strong sulphuric acid; it is then thoroughly washed and dried at a low temperature.

The cotton is not changed in appearance other than becoming somewhat wrinkled. When well prepared it burns completely, leaving no residue. The temperature at which it takes fire varies from 100° to 180° according to the manner in which it has been prepared. It is cellulose in which from six to nine atoms hydrogen have been replaced by an equivalent quantity of the monad radicle NO_2 that, having the formula $C_{18}H_{21}O_{15}9NO_2$, has the greatest explosive energy. Pyroxylin regenerates cellulose in contact with ferrous chloride. If cellulose be considered a sort of alcohol, as claimed by some, pyroxylin would be a nitric ether of this alcohol.

Pyroxylin has the advantage over gunpowder of

being more easily prepared, and of remaining unaffected by moisture, but its cost is relatively greater, and its shattering power renders its employment dangerous.

The term *collodion* (from κολλα, glue) is given to a preparation obtained by dissolved gun-cotton in a mixture of 1 part of alcohol and 4 parts of ether.

Chas. H. Mitchell has made (52–74–235) a number of experiments, with the view of ascertaining the relative proportions of cotton and acid, together with the proper time of maceration necessary to produce a cotton which should combine the largest yield with the highest explosive power and solubility.

The following formula was at length adopted:

Raw cotton,	2 parts.
Potassium carbonate,	1 "
Distilled water,	100 "

Boil for several hours, adding water to keep up the measure; then wash until free from any alkali, and dry. Then take of—

Purified cotton,	7 oz. av.
Nitrous acid (nitric, saturated with nitrous acid), s. g. 1.42,	4 pints.
Sulphuric acid, s. g. 1.84,	4 "

Mix the acids in a stone jar capable of holding 2 gals., and when cooled to about 80° Fahr., immerse the cotton in small portions at a time; cover the jar and allow to stand 4 days in a moderately cool place (temp. 50° to 70° Fahr.) then wash the cotton in small portions, in hot water, to remove the principal part of the acid; pack in a conical glass percolator, and pour on distilled water until the washings are not affected by solution of barium chloride.

Collodion, on spontaneously evaporating, forms a transparent and impermeable membraneous coating, and is much employed in photography, also somewhat in surgery.

Cellulose is attacked by chlorine; the use of solutions of chloride of lime, and of chlorine, in large quantities in washing, or bleaching, will cause a rapid deterioration of linen or cotton goods.

Schweizer has shown that cotton, paper, etc., is very easily dissolved by an ammoniacal solution of copper. Attempts by the author to employ this solution for a "water-proof" coating of fabrics, as has been suggested, failed to yield a satisfactory result, on account of the liability of the coating to crack and peel off.

Péligot has found in the skin of silk worms, and Schmidt has discovered in the envelopes of the Tunicates, a substance, *tunicine*, which has the composition and properties of cellulose.

Linen, hemp, cotton, wood and paper are all essentially cellulose.

AMYLACEOUS SUBSTANCES.

These substances are almost universally present in plants; particularly that known as *starch* or *fecula*.

The potato yields about 20 per cent. of starch. In order to obtain it, this root is grated and the pulp placed upon sieves, arranged one above the other, and through which a stream of water flows.

The grains of starch being extremely minute pass through the meshes of the sieve, while the walls of the cells remain behind. The starch is washed, drained, and dried, first at ordinary temperature, afterwards by the application of a moderate heat.

STARCH. $x(C_6H_{10}O_5)$ probably $C_{18}H_{30}O_{15}$. Flour contains, besides starch, nitrogenous substances, denominated gluten; this gluten is capable of fermenting, whereupon it becomes soluble, while the starch remains unaltered and insoluble. Under these conditions the gluten gradually dissolves, disengaging ammoniacal compounds, hydrogen sulphide and other products of putrefaction.

At the end of twenty or thirty days, the gluten having become dissolved, the liquid is removed, and the starch, washed and dried, shrinks into columnar fragments, which are readily pulverized by gentle pressure.

STARCH. 211

A more modern method is that employed in France, which is essentially the same as the process cited above, as that used in making potato starch here. The water carries away the starch while the gluten remains behind in the form of an elastic mass, which is also utilized. For this purpose it is incorporated with flour poor in gluten, to be made into macaroni, and for the manufacture of a very nutritive preparation, " granulated gluten;" it is also employed, according to the recommendation of Bouchardat, in making bread for persons afflicted with diabetes.

Starch, examined with a microscope, exhibits flattened ovate granules of different size in various plants, but always very small. Those of the Rohan potato have a length of 0.185 mm.; the smallest are those of the *Chenopodium quinoa* whose length is 0.002 mm.

When starch is heated with water to 70°, the granules increase from 20 to 30 times their original volume, and become converted into a tenacious paste. A small quantity of the starch passes into solution, and to this the name *amidin* has been given. Starch paste and the solutions of starch have the characteristic property of becoming blue in contact with small quantities of iodine. The liquid becomes colorless at about 70°, but regains its color on cooling. If to this blue liquid a solution of a salt, sodium sulphate for instance, be added, we obtain a dark-blue floculent precipitate. This substance, called starch iodide, is not a chemical compound, but a sort of lake, containing variable quantities of iodine diffused throughout the starch and solvent. This reaction with iodine is a very valuable test for starch, but is open to several fallacies, and apt to mislead in inexperienced hands.

Until lately, it has been claimed that starch is insoluble in water, and that if water in which starch has been boiled gives with iodine the characteristic reaction of this substance, it is due to particles of starch sufficiently minute to pass through the pores of the filter. But the results of the experiments of Maschke and Thenard, show that if starch is heated for some time at 100°, it is partially transformed into a variety soluble in water. This substance is colored by iodine; it furnishes, on evaporation, a gummy solid which is precipitated by alcohol as an amorphous powder.

If we boil starch for a long time with water it is converted into a substance called *dextrin*. The presence of a small per centage of sulphuric acid facilitates this change, which is soon followed by the transformation of the dextrin into glucose. The sulphuric acid is not at all altered during the reaction.

The change of starch into glucose also takes place when water containing starch, and to which germinated barley has been added, is heated to about 70°.

This transformation is due to a substance called *diastase* (from διαστασις, separation), which is formed in the seed during germination. The production of diastase on the formation of the young shoot, explains how starch becomes soluble and serves as nutriment to the young plant.

The *ptyalin* of the saliva, the pancreatic juice, the

soluble parts of beer yeast, gluten, and many other substances, are capable of producing this transformation of starch into dextrin and glucose.

It has generally been considered that the molecule of starch, in being transformed into glucose, simply united with one molecule of water directly, thus:

$$C_6H_{10}O_5 + H_2O = C_6H_{12}O_6.$$

Musculus, however, claims to have established that the starch is first transformed into a soluble metamer, and this, thereupon, splits up into dextrin and glucose:

$$C_{18}H_{30}O_{15} + H_2O = \underbrace{2C_6H_{10}O_5}_{\text{Dextrin.}} + \underbrace{C_6H_{12}O_6}_{\text{Glucose.}}$$

By further action, the whole of the dextrine becomes converted into glucose, (2–[3]60–203).

Starch, heated simply to about 160°, is also changed into dextrin.

It is attacked by dilute nitric acid, nitrous vapors are given off and different substances are produced, chiefly, however, oxalic acid.

If starch is agitated with fuming nitric acid, it is dissolved and water precipitates from the solution a nitrous compound which is explosive.

The alkalies, in concentrated solutions, when heated with starch disorganize and dissolve it. Solutions containing two to three per cent. of alkali, accelerate the formation of starch paste.

Starch is employed in the laundry and therapeutically in poultices, injections and baths.

Tapioca is the starch of the root of the *Jatropa manihot*, called cassava or manioc.

Sago is obtained from the pith of various sago palms.

Arrow-root is the starch of the *Maranta arundinaceæ*, and one or two other tropical plants.

Salep is obtained from the *Orchis mascula*.

INULIN. There has been found in the roots of the Jerusalem artichoke, of the chicory, and the bulbs of the dahlia, a substance isomeric with starch, called *inulin*.

LICHENIN. There is extracted from certain lichens and mosses a substance called *lichenin*, which has the property of swelling in cold water and of being dissolved in boiling water. It is prepared by treating Iceland moss with ether, alcohol, a weak solution of potassa, and finally with dilute hydrochloric acid.

There exists in the animal organism a variety of starch designated by the name of *glycogen*.

DEXTRIN, OR DEXTRINE.

$$C_6H_{10}O_5.$$

To prepare dextrin, starch may be heated with water containing a small quantity of sulphuric or oxalic acid; the operation should be arrested when the liquid gives with iodine only a wine-colored reaction.

FLOUR. 215

For the acids, a small quantity of germinated barley may be substituted, placed in a bag immersed in the liquid. Dextrin thus prepared always contains glucose. It may be obtained free from this substance by heating starch with $\frac{1}{8}$ its weight of water and $\frac{3}{1000}$ of nitric acid.

Dextrin is amorphous, slightly yellow, very soluble in water, insoluble in alcohol and concentrated ether.

It is used somewhat in preparing bandages in case of fracture, and very extensively as a paste for calico-printers.

Dextrin, forms viscid adhesive solutions which are used for the same purposes as gum-arabic. The mucilage used by the U. S. government for postage stamps is composed of dextrin two ounces, acetic acid one ounce, water five ounces, alcohol one ounce. Dextrin may be distinguished from gum-arabic by not being precipitated on adding a dilute solution of lead acetate, and by furnishing with nitric acid a solution of oxalic acid and not a precipitate of mucic acid.

FLOUR.

Amylaceous substances are of great importance as food. Wheat and other cereals are the most important sources of these aliments.

Starch, as also sugar and the neutral carbohydrates, are respiratory foods whose principal effect is the production of heat by being oxidized, or burned, in the body.

The composition of four of the leading cereals is herewith given:

	Water.	Starch.	Dextrin and Glucose.	Cellulose.	Nitrogenous Substances.	Fat.	Mineral Substance.
Wheat,	14.0	59.5	7	1.7	14	1.2	1.5
Rye,	16.0	57.5	10	3.0	9	2.0	2.0
Oats,	14.0	53.5	8	4.0	12	5.5	4.0
Rice,	14.5	77.0		0.5	7	0.5	0.7

The sticky, elastic substance found with starch in flour is *gluten* (called also *glutin*), and is a mixture of various proximate compounds, but chiefly of three; legumin, or vegetable casein, fibrin and gelatine.

Flour of good quality is dry and soft to the touch; it forms with water an elastic, non-adhesive dough.

The value of flour depends largely upon the gluten it contains, though not as stated in most authors upon the percentage of this substance, but upon the quality rather, as shown by recent investigations of R. W. Kunis (26-74-1487).

The modern "patent process," originating in Minnesota, is mainly a method of grinding which introduces into the flour more gluten than in older processes.

GUM.

$$C_6H_{10}O_5.$$

This substance is very widely distributed in the vegetable kingdom. Gums either swell in water or are dissolved, imparting to it a mucilaginous consistency.

From a chemical standpoint they are essentially characterized by giving a precipitate of *mucic acid* on being boiled with nitric acid, and by precipitating lead subacetate.

GUM-ARABIC, ARABIN. This gum exudes from different species of acacias, as *Acacia arabica, A. senegalensis, A. vera;* it is obtained from Arabia and Senegal.

According to Fremy, gum-arabic is a salt formed by the combination of an acid, *gummic* or *arabic acid*, with lime and potassa. This acid may be isolated by pouring hydrochloric acid into a solution of gum, and adding alcohol; an amphorous deposit is formed which, dried at 120°, has the formula $C_6H_{10}O_5$. This acid is very soluble in water. Its solution is levogyrate, like that of gum-arabic. On being heated to 150° it is transformed into a substance insoluble in water called *meta-gummic acid*, whose salts are likewise insoluble. Gum-arabic gives with ferric salts an orange-colored, floculent precipitate soluble in acids.

CERASIN. The gum which exudes from cherry and plum trees is a mixture of soluble gummates and insoluble meta-gummates; hence it is only partially soluble in water.

Cerasin becomes soluble on being boiled with water, as the meta-gummates are transformed into gummates by the action of boiling water.

These gums heated with dilute sulphuric acid furnish a dextrogyrate sugar.

Gum-tragacanth often contains starch.

MUCILAGE OR BASSORIN. There exists in the seeds of the quince and flax, in the roots of the marsh-mallow and in portions of many other plants, a substance or substances, which, exposed to the action of boiling water, furnish a thick mucilage, which appears to consist of a soluble, together with an insoluble substance. Nitric acid converts this mucilage into mucic and oxalic acids. Gum and mucilage are frequently employed as emollients, and in syrups, also extensively in confectionery.

PECTIN GROUP. Many roots, as the carrot, beet, etc., also green fruits, contain a neutral gelatinous substance, insoluble in water, alcohol and ether, called *pectose*. It is that which gives to green fruits their harshness. This substance is modified during the ripening of the fruit and becomes soluble, vegetable jelly, or *pectin* (from πηκτίς, a jelly), to which Fremy assigns the formula $C_{32}H_{48}O_{32}$.

Pectin, submitted to the action of a ferment found in the cellular tissues of vegetables, called *pectase*, or of cold, very dilute, alkaline solutions, is changed into a gelatinous acid called *pectosic acid*, then into another substance likewise gelatinous, which is known by the name of *pectic acid*. All these substances are amorphous, and non-nitrogenous. Their formulæ are not yet definitely determined.

According to Fremy, to whom we are indebted for the foregoing facts, the jelly obtained from the current and other fruits is due to the action of the pectase on the pectin of these fruits.

These substances resemble gums in producing, on boiling with nitric acid, a precipitate of mucic acid.

Much doubt still exists respecting the composition of the pectin group.

LEGUMIN OR VEGETABLE CASEIN.

Legumin is found in most leguminous seeds, such as sweet and bitter almonds, also in beans, peas, etc., the latter containing about 25 per cent. It is considered to be identical with casein by Liebig and Woehler.

It may be obtained by digesting coarsely powdered peas in cold or tepid water for two hours, allowing the starch and fibrous matter to subside, and then filtering the liquid. It forms a clear, viscid solution, which is not coagulated by heat unless albumen is also present, but, like emulsin and unlike albumen, it is precipitated by acetic acid. It is coagulated by lactic acid, also by alcohol; in the latter case the precipitate is redissolved by water.

Acetic acid, diluted with 8 to 10 parts of water, is carefully dropped into the filtered solution obtained above, and the legumin is precipitated; an excess of the acid should be avoided, as this would dissolve the precipitate. It falls in the shape of white flakes, and after having been washed on a filter should be dried, pulverized and freed from adhering fat by digestion in ether. Legumin may be obtained from lentils with the same facility as from peas; but it is less easily procured from beans (haricots), in consequence of their containing a gummy matter which interferes with its precipitation and with the filtration of the liquids.

The chemical properties of legumin are identical with those of casein.

Liebig supposes that grape-juice and other vegetable juices which are deficient in albumen, derive their fermentation power from soluble legumin. This principle is soluble in tartaric acid, and to its presence he ascribes the tendency of sugar to form alcohol and carbon dioxide instead of mucilage and lactic acid.

VEGETABLE ALBUMEN.

Vegetable albumen is contained in many plant-juices and is deposited in flocculi on applying heat to such liquids. It can also be precipitated by nitric acid, tannin and mercuric chloride precisely like animal albumen. Vegetable albumen is composed of carbon, hydrogen, nitrogen, oxygen and sulphur. There is no trustworthy formula for this substance.

ANIMAL CHEMISTRY.

ANIMAL CHEMISTRY.

THE substances serving as materials to build up the structure of animals are of a varied nature; they may, however, be grouped into four classes:

    I. FARINACEOUS AND SACCHARINE.
    II. FATTY.
    III. NITROGENOUS.
    IV. MINERAL.

We have already studied the first, second, and fourth of these classes; we will now proceed to examine those of the third.

NITROGENOUS SUBSTANCES.

It is generally considered that these substances act a different part in the organism from that of the saccharine and fatty bodies, these latter serving exclusively as heat producers, and being decomposed and ultimately consumed (oxidized) in the respiratory process, have therefore received the name of *respiratory* foods. The nitrogenous principles (*albumen, casein, fibrin,* etc.) serving to form the tissues have, likewise, received the denomination *plastic* foods. The distinction thus made is too restricted, as we shall show later.

Dumas and Cahours have proven that the cereals and other plants employed as food contain similar principles to those found in flesh, and especially that albuminoid matter exists in plants as well as animals.

The albumen of the blood and that of wheat are alike. In the gluten of wheat albuminoid substances are found which are hardly distinguishable from animal albumen, fibrin, and casein.

These substances are characterized:

1st. By their amorphous structure. The three substances mentioned never crystallize; and as they are also non-volatile, it is difficult to form an idea of their constitution, and represent them by a formula. This formula must necessarily be very complex, as sulphur forms a constituent, though present only in very small quantity. Lieberkühn represents their composition by the expression

$$C_{72}H_{112}N_{18}SO_{22}.$$

2nd. By their extreme instability. The apparently most insignificant circumstance causes them to pass from a soluble to an insoluble condition, or *vice versâ*, and produces their transformation. They are decomposed with great facility under the action of air and water. This very exceptional instability constitutes a property of the greatest interest, as it permits these substances to take part in a wonderful manner in the varied transformations which occur in living organisms, and it might be said that they are the principal agents of development in animals and plants. We shall presently see that, whatever this real albuminoid substance may be, it is transformed in the stomach into identical substances—*peptones;* also, that during the incubation of the egg the albumen is seemingly changed into fibrin.

CLASSIFICATION —The albuminoid substances are very numerous, and may be classed into two groups. Those of the first group contain:

Carbon	53.5
Hydrogen	6.9
Nitrogen	15.6
Oxygen	24.0
	100.0

They contain, besides, 0.4 to 0.5 per cent. of sulphur, unlike those of the second group, which usually contain no sulphur. In addition, they often contain small quantities of mineral substances.

The first are more specially designated by the name of *albuminoid* substances, as albumen is the most characteristic member of the group. They are also known by the name of protein substances, because Mulder claimed they might be considered as formed of a single radical *protein*, to which are united variable proportions of sulphur, phosphorus, etc.

The principal members of this group are: *albumen*, of which several modifications are recognized—the *paralbumen, metalbumen*, etc.; *fibrin*, of which there are several kinds—the *fibrin* of the blood, *fibrin* of the muscles or myosin; *casein*, regarded by some as a combination of albumen and alkali; *hemoglobin* or *hematocrystallin*, the colouring matter of the blood, which is distinguished from most other albuminoid substances by its property of crystallizing; *vitellin*, the principle of the yolk of an egg; also, several principles, *icthin, icthlin* (ἰχθύς, a fish), *emydin*, the first two obtained by Valenciennes and Frémy from fishes' eggs, the latter from the eggs of the turtle.

The composition of these substances is identical or very similar; a formula cannot be given with precision.

The substances of the second group generally contain less sulphur, often none, and appear to be derived from the first by the addition of nitrogen and oxygen. They contain in per cent.:

Carbon	50.0
Hydrogen	6.3
Nitrogen	16.8
Oxygen	26.6
	100.0

In this group we find—*ossein*, the organic substance of bones, which is converted into gelatin by the action of boiling water; cartilage, a substance very analogous to the latter, and which is transformed by boiling water into *chondrin;* various principles concerned in the digestive phenomena, as the *ptyalin* of the saliva, the *pepsin* of the gastric juice, the *mucin* of the mucus, the *pyin* of pus, etc.; together with different products which result from the action of the gastric juice upon nitrogenous substances, and which are called *albuminoses* or *peptones*.

GENERAL CHARACTERISTICS.—The substances of these two groups on being heated give off an odour of burnt feathers. On distillation they produce water, empyreumatic oils, and ammonium carbonate, sulphide, and cyanide. Carbon remains in the retort.

The substances of the first group, on being heated to $50°-60°$ with a solution of potassium hydrate, lose their sulphur and are dissolved. If we add acetic acid to this liquid, dark grey flakes of a substance (protein of Mulder) are thrown down. The substances of the second group do not possess this property. On protracted boiling with a caustic alkali they yield:

Tyrosin	$C_9H_{11}NO_3$
Leucin	$C_6H_{13}NO_2$
Glycocol	$C_2H_5NO_2$

Some are soluble, others insoluble, in water; they are, in general, insoluble in alcohol, ether, and chloroform.

Hydrochloric acid diluted with 1,000 times its weight of water dissolves some, a few swell up simply; upon others it has no effect. Hot concentrated hydrochloric acid attacks all these substances, and the resulting products are the same as those which are obtained (and more readily) with sulphuric acid. These products are chiefly glycocol, leucin, and tyrosin. Nitric acid colours them yellow (xanthoproteic acid). Ordinary phosphoric and acetic acids do not precipitate the substances of the second group, but redissolve them even when coagulated.

Solutions of albuminoid substances in potassium hydrate do not precipitate copper salts. Heated with oxidizing reagents, as a mixture of potassium bichromate and sulphuric acid, they furnish several members of the series of fatty acids, and the aldehyds corresponding to these acids. The albuminoid substances are decomposed during the process of respiration in the same manner as when under the action of oxidizing agents.

Ammoniacal solutions of copper dissolve albuminoid substances as they dissolve cellulose, which fact would seem to connect the albuminoid substances with cellulose, and to give certain weight to a theory of Hunt, which considers the albuminoid substances as cellulose which has combined with the elements of ammonia and parted with the elements of water.

ALBUMEN.

This substance is found both in vegetable organisms (cereals) and in animal organisms (serum of the blood, white of egg, lymph, chyle).

Wurtz obtains it by mixing white of eggs with twice its weight of water, straining and precipitating the albumen with a solution of lead acetate. The precipitate is washed with cold water and decomposed with a current of carbon dioxide, which precipitates the lead, while the albumen remains in solution. If this liquid be evaporated at a temperature below 59°, it is deposited in a soluble state; if a quantity be heated to 63°, a portion of the albumen is coagulated; but if the temperature is not raised above 74°, four-fifths remain dissolved; consequently it would seem as though there were several kinds of albumen, but the nature and amount of foreign substances present are the principal causes of these differences. If the solution is very dilute, coagulation will not take place. Heating is not the only mode of producing this change; alcohol, acids—with the exception of a few, such as hydrogen phosphate, H_3PO_4, hydrogen tartrate and hydrogen acetate—the metallic salts, creosote, tannin, etc., also effect it. The alkalies prevent this action. Gautier obtains albumen by dialysis.

Soluble albumen is without odour, and is more soluble in saline than in pure water. Very dilute hydrogen chloride precipitates solutions of albumen, the precipitate being redissolved by an excess of the acid. This solution does not contain albumen, but a substance probably isomeric with it, which is, however, more easily obtained from muscular tissue: it is called *syntonin*.

Among the products of the putrefaction of albumen,

Nencki (18-'78-71) has obtained butyric and valerianic acids.

Insoluble albumen heated with water in a sealed tube to 150° or 160°, dissolves, but this modification is not coagulated again by heat.

Animal albumen containing 1.5 per cent. of soda may be regarded as a weak acid, and in presence of alkaline solutions it dissolves. A few drops of potassium hydrate are sufficient to form with albumen a gelatinous compound, called potassium albuminate, which is soluble in water and no longer coagulable by heat. This liquid, diluted with water, is rendered turbid by acetic acid, but the precipitate is redissolved by an excess of acid.

Albumen of the Serum.—This is easily soluble in concentrated hydrochloric acid, and is not precipitated by ether. Injected into the veins it is absorbed.

Egg Albumen.—This is more difficultly soluble in concentrated hydrochloric acid, and is precipitated by ether. Injected into the veins it is absorbed in very minute quantities, and can be found again in the urine.

Albuminoid substances (fibrino-plastic substance, fibrinogene) are found in the blood; they have the general characteristics of albumen, but are distinguished from it by being precipitated with carbon dioxide. The soluble matter of the crystalline lens of the eye also possesses this property.

The coagulation of albumen by alcohol, tannin, and heat, and the consequent formation of a sort of network which fills the whole liquid, and which precipitates all matters held in suspension, as well as certain substances in solution, explains the employment of white of egg for the clarifying of wine, syrups, also as a mordant, and the use of blood in sugar refining.

FIBRIN.

The blood of animals coagulates spontaneously shortly after leaving the body. This is due to the solidification of a substance called fibrin, which, on solidifying, forms a sort of net-work, imprisoning the globules of the blood, and gives rise to a gelatinous mass (clot). The researches made to explain this process of coagulation will be mentioned further on. Ether accelerates this coagulation. Sodium sulphate and glycerine retard or even arrest it. Pure fibrin may be obtained by beating fresh blood with twigs. It attaches itself to the twigs, and if then washed with water, and afterwards with alcohol, we obtain a dark-grey filamentous substance, which is insoluble fibrin. It may also be obtained by working clotted blood in water as long as it colours the water.

Fibrin is insoluble in water, hot or cold, but if heated with it in close vessels it gradually loses its property of solidifying. It is soluble in alkaline solutions, and precipitable again by acids.

Lehman has concluded from his analyses that fibrin is oxidized albumen. Smée affirms that if oxygen be passed into defibrinated serum, heated to about

36, the albumen is gradually transformed into flocks of fibrin. This subject needs further investigation.

Fibrin of blood swells when treated with water containing 1-1000th part of hydrochloric acid. It is dissolved in stronger hydrochloric acid, and is then converted into syntonin. Freshly precipitated fibrin is dissolved at 35° to 40° in water containing certain salts, and notably in that containing potassium nitrate or sodium sulphate; it decomposes hydrogen peroxide.

The albumen in the egg is transformed into fibrin during incubation; inversely, if fibrin be kept under water, it gradually becomes soluble, and this liquid, like albumen, is coagulated by the action of heat.

Varieties of Fibrin.—The gluten which constitutes the plastic substance of cereals, has the composition and general properties of fibrin.

When well-washed, muscular tissue is macerated with water containing ten parts in 100 of sea salt, it is partially dissolved; if this solution is poured into water a gelatinous mass is obtained on agitation. This substance washed on a filter has received the name of *myosin*, or *musculin*.

It is soluble in acids, in dilute alkalies, and in a solution of sea salt; this last solution coagulates at about 60°.

Myosin, on dissolving in dilute acids, is changed into *syntonin*, which, like myosin, is soluble in acids and alkalies, but from which it is distinguished by its insolubility in salt water. Syntonin is more easily obtained by macerating flesh, which has been completely deprived of blood by prolonged washing with water, containing 0.01 of hydrogen chloride. The macerated flesh is almost entirely dissolved This solution is filtered and exactly neutralized with sodium carbonate; the syntonin is precipitated in a grey, flocculent form.

Blood fibrin contains about: $C = 52.6$; $H = 7.0$; $N = 16.6$; $S = 1.2$ to 1.6; $O =$ the difference, authors not agreeing very closely as to its exact composition.

CASEIN.

Casein is the nitrogenous principle of milk. To extract it, milk is brought to boiling, and a few drops of acetic acid added. An abundant coagulum of casein mixed with butter (caseum) is formed. The pure casein is separated by washing this coagulum several times with water, alcohol, and ether.

Casein is difficultly soluble in water, but is dissolved by alkalies. It forms with the alkalies soluble compounds, and with the other bases insoluble salts. Casein has the composition of the albuminates of soda, differing, however, from these by various reactions, and by the amount of its levogyrate action on the polarized ray of light.

Solutions of casein are not coagulated by heat, they simply become covered with a white film. They are precipitated by acetic and other organic acids; milk curdles spontaneously, on account of the lactic acid formed in it.

Many substances such as tannin, alcohol, plants with acid reactions and several others, the flowers of the artichoke, of the thistle, of the butterwort (*Pinguicula vulgaris*), and, above all, rennet from the stomach of a sucking calf, cause coagulation in milk.

LEGUMIN, OR VEGETABLE CASEIN.

Braconnot extracted, by means of water, from the seeds of leguminous plants (beans, peas) a substance called *legumin*, and which has a close analogy to casein.

VITELLIN.

This substance is prepared by treating boiled yolk of egg with ether, which extracts the fatty matters. There remains a white substance insoluble in water. It can be obtained in a soluble state by mixing fresh yolk of egg with water. The clear liquid coagulates at about 70°, like albumen, of which it possesses the general properties.

OSSEIN, GELATIN, CHONDRIN.

The compounds of this second group are probably formed of a single substance, whose elements are differently aggregated, and also mixed with variable quantities of mineral substances. They are insoluble in water, alcohol, and acetic acid; they swell in cold and dissolve in hot alkaline solutions.

The organic substance of bones (*ossein*), treated with

boiling water, furnishes *gelatin*. The cartilages, under the same circumstances, furnish a product which has most of the properties of gelatin, but which differs from it in being precipitated by acids and by alum; it is called *chondrin*.

To prepare gelatin, bones are treated with boiling water to remove the grease, then macerated with water acidulated with hydrochloric acid, which dissolves the mineral portions (calcium carbonate and phosphate). The organic portions remain undissolved, retaining the form of the bone, yet flexible and elastic. The solution is poured off and employed in the manufacture of calcium hypophosphite, or of composts. The organic substance, well freed from acid by washing in milk of lime or a weak solution of sodium carbonate, is put into boilers with water, which is gradually raised to the boiling point. The organic matter gradually enters into solution. It is now decanted into a vat heated over a water bath, where various undissolved substances are deposited, and whence it is drawn into wooden moulds, where it solidifies. The gelatin is removed from the moulds, cut into thin slices, dried on nets, and is now the *glue* of commerce.

The tendons, skin, horns, and clippings of hides are also employed for the manufacture of glue; they are simply treated with boiling water.

Darcet showed in 1817 that gelatin could be made directly from bones by digesting them with steam heated to 104°. The solution obtained has the appearance of soup, and it was hoped to thus produce a very substantial nutriment quite cheaply; but it has been found that the nutritive power of this substance is very small, and the use of "*gelatine food*" has been abandoned.

The purest gelatin is the ichthyocol, or *isinglass*. It is made chiefly in Moldavia and on the borders of the Caspian Sea, from the swimming bladders of the sturgeon and of the *acipenseres*.

Pure gelatin is solid, colourless, and transparent. Boiling water dissolves it in large quantities. The liquid solidifies to a jelly on cooling: one per cent. is sufficient to give water a gelatinous consistency. Continued boiling with water deprives it of the property of solidifying in the cold, or gelatinizing. Boiled with dilute sulphuric acid, it is transformed into glycocol.

DIGESTION.

An organized being cannot live without nourishment, that is, without obtaining from the bodies which surround it the materials necessary for the formation and the metamorphoses of its tissues.

The food of animals is rarely assimilable in the state in which it is found in nature; therefore it must undergo a preparation which shall render it absorbable. Hence the existence of a particular function, *digestion*. This function is performed by the *digestive organs*. They vary in complexity with different animals; they differ in form according to the nature of the food.

In man the digestive apparatus is very complex. If the food is solid it must be dissolved. Every liquid, however, is not immediately assimilable; it often also must be transformed in chemical and physical character. We shall now follow the food through the process of digestion, and explain the manner in which each class of aliments becomes soluble and absorbed.

SALIVA.

In the mouth the food is subjected to mechanical action under the influence of a liquid secreted by glands situated in pairs on each side of the mouth (parotid, submaxillary, and sublingual).

Tubes have been introduced into the ducts of the parotid and submaxillary glands, and by exciting secretion, the products of these glands have been separately examined. The salivas are not alike, and have different digestive properties, their combination, mingled with mucus, constituting "mixed saliva."

The parotid secretion is a clear liquid, not viscous, and slightly alkaline, containing 1.0 to 1.6 per cent. of solid substances, among which are alkaline chlorides and phosphates; an organic substance soluble in alcohol and water; another, *ptyalin*, which is the most important principle of the saliva; and finally, potassium sulphocyanide.

Ptyalin contains potassium, sodium, and calcium. It resembles compounds of albumen with these bases, is, however, gelatinous and not coagulable by heat or by most metallic salts. It is precipitated by mercury bichloride, lead acetate, and tannin.

The submaxillary glands are dependent upon the chorda tympani nerve, and the branches of the great sympathetic nerve. The secretion varies as it is excited by the one or the other of these nerves.

The liquid secreted after an excitement of the great sympathetic nerve is thick, alkaline, and rich in solid substances. The liquid obtained by the excitement of the chorda tympani is less concentrated. It is alkaline, and contains epithelial cells, small quantities of albumen, globulin, and a substance (mucin) to which its mucilaginous appearance is due, and which is not found in the parotid secretion.

The liquid of the sublingual glands has not as yet been obtained pure; concerning it we know only that it is a viscous solution.

Buccal mucus has a slight acid reaction.

Mixed saliva is a turbid, ropy, inodorous, tasteless liquid. It deposits *débris* of epithelium. In man its density varies from 1.007 to 1.008.

It has an alkaline reaction, and contains from 0.7 to 1.0 per cent. of solid substances, of which about one-third is inorganic, chiefly alkaline carbonates, phosphates, and chlorides. It contains in solution more carbon dioxide than even venous blood.

1,000 parts of saliva contain :—

	Mitscherlich.	Jacubowitsch.
Water	984.50	992.16
Solid substances	10.50	4.84
Ptyalin	5.25	1.34
Mucus and epithelium	0.05	1.62
Sulphocyanogen	..	0.06

According to Longet, potassium sulphocyanide is a normal product of the saliva. It is recognized by placing in the saliva a ferric salt, which is coloured red. This salt does not exist in the blood, perspiration, lacrymal fluid, or pancreatic juice. Its amount is always very small, and its presence in saliva is doubted by Gautier.

On boiling saliva it becomes opalescent, on account of the precipitation of albumen. Nitric acid colours it yellow by attacking the albuminoid substances. Alcohol precipitates from it ptyalin, mixed with nitrogenous compounds.

Saliva exposed to the air becomes covered with a film of calcium carbonate, and concretions of this substance are often found in the salivary ducts and on the teeth.

An adult can secrete about 1,200 to 1,500 grains of saliva in twenty-four hours: the actual quantity varies with the dryness of the food.

The saliva possesses evident mechanical functions in digestion. It facilitates mastication by impregnating the food; it lubricates the bolus, and renders deglutition possible; and finally, by virtue of its viscous and frothy consistency, it imprisons air, which passes into the œsophagus with the food.

Deglutition is favoured much more by the mucus than by the saliva proper; this mucus is secreted by glands found in the walls of the mouth and pharynx.

It has been contested that the saliva has for a chemical function the saccharification of starch, as the food does not remain but for an instant in contact with it, and as the amount of saliva secreted is independent of the amount of starch in the food. The proportion of saliva increases when the food is dry or hard, and diminishes when it is soft, even when it is formed of boiled starch: in short, it seems to vary inversely with the humidity of the food.

It has been remarked that salivary glands exist in a rudimentary state in animals which do not masticate their food.

Mialhe indicates the following experiment: Chew some unleavened bread, then place it on Berzelius test paper. Rub another portion of the same bread with water, and filter the liquid. The first is not coloured by iodine, and becomes brown on being boiled with potassa. The second turns blue with iodine.

According to the same chemist this action is due to ptyalin, an amorphous substance insoluble in alcohol, of which 1 to 2 per cent. is present in the saliva, and which is able to saccharify as much as 2,000 times its own weight of starch; it also effects this change with extreme rapidity. This substance has then the property of vegetable diastase.

It has also been shown directly by Messrs. Mialhe, Longet, and Schiff, that even if pure gastric juice itself does not have the property of saccharifying starch, this saccharification by the saliva is not arrested by the acidity of the gastric juice, and consequently the saliva which is carried into the stomach can continue to saccharify the starchy food in this organ. It is then very probable that the saliva performs this service in the process of digestion, though some claim that this action is only on food not yet thoroughly mixed with gastric juice. It appears to have no action on sugar, gum, cellulose, or albuminoid compounds.

In inflammatory diseases of the mouth, as in the thrush, the saliva becomes acid, and weak alkaline beverages are prescribed. In Bright's disease urea is found in the saliva. Mercury is also present in cases of mercurial salivation. After the use of preparations containing iodine and bromine, these substances are found in this secretion.

The tartar of the teeth contains, in 100 parts, 25 of organic substances, 75 of inorganic substances, formed chiefly of calcium phosphate; the remainder is calcium carbonate, iron, and silica.

GASTRIC JUICE.

The gastric juice is secreted in the mucous membrane of the stomach by an immense number of glandular follicles, though not secreted when the stomach is empty.

As soon as food enters the stomach, the mucous membrane swells, assumes a blood-red colour, and the gastric juice is at once secreted.

The secretion can also be excited by irritating this membrane by ice, cold water, wine, gall, coffee, bismuth subnitrate, sodium bicarbonate, and alkaline substances in general. According to L. Corvisart, gastric juice secreted by mechanical irritation is most rich in digestive principles. We can easily procure gastric juice, or rather a mixed liquid, formed of this juice, stomachic mucus, and saliva, by making an aperture in the stomach of an animal, and it may be obtained free from saliva by previously ligating the œsophagus.

The gastric juice may be freed, to a great extent, from the mucus by filtration.

The mucus is alkaline. When the stomach has not received food for a long time, the mucous secretion alone is produced, and the liquid in the stomach may become alkaline. The gastric juice forms an almost colourless liquid, with a faint odour, decidedly acid reaction, and is somewhat denser than water. It may be preserved for a very long time without alteration. It loses its digestive properties on ebullition, but is not altered by cold.

Schmidt obtained the following analysis of gastric juice mixed with saliva:—

	Man.	Dog.
Organic matter	8.79	17.12
Hydrochloric acid	0.20	3.05
Potassium chloride	0.55	1.12
Sodium ,,	1.46	2.50
Calcium ,,	0.06	.62
Ammonium ,,		.47
Calcium phosphate		1.73
Magnesium ,,	0.14	0.23
Iron ,,		0.08
Water	988.80	973.08
	1000.00	1000.00

These analyses are the mean of several. It is, moreover, very evident that the composition of this liquid, as well as that of others in the body, must vary, not only in the quantity, but sometimes even in the nature of their constituents.

All analyses of the gastric juice show that it is an acid substance. The nature of this acid has been the object of much discussion between different experimenters, and principally between Messrs. Blondlot and Schmidt. According to the first, the acidity is due to acid calcium phosphate; according to the second, to hydrochloric acid.

It is true that calcium phosphate is found in the gastric juice, but according to Lehmann and Schiff, this salt is formed by the action of the gastric juice on the substance of the bones, and does not exist in the gastric juice of animals deprived of this food. Consequently this substance is not a normal and constant constituent of gastric juice.

Schmidt points to the following experiment. He determines the amount of chlorine in a known weight of gastric juice, by means of silver nitrate, and also determines the bases. Now, the quantity of hydrochloric acid corresponding to the weight of chlorine determined by analysis, is always more than sufficient to neutralize the bases found; hence he concludes that hydrochloric acid exists in a free state in the gastric juice.

Moreover, having determined the amount of free acid by saturating the gastric juice with a standard solution of a base, he found that the amount of free acid was about equal to the weight of the excess of hydrochloric acid resulting from the previous determination.

Lactic acid is generally found in the gastric juice, and, according to Messrs. Bernard and Barreswill, this is probably the acidifying principle; according to others, it exists there only when starchy food is used. Thus, recently, Rabuteau (9–80–61), by neutralizing gastric juice with quinia evaporating, treating with amyl-alcohol, and obtaining the crystallized quinia, salt dissolved in the amyl-alcohol, found the free acid of the stomach to be hydrochloric acid; he was not able to discover lactic acid in the gastric juice.

Butyric and acetic acids have also been detected in the gastric juice.

When lactic acid is distilled with a dilute solution of a chloride, hydrochloric acid is found in the product: possibly the hydrochloric acid is produced in the stomach by the reaction of the lactic acid on the alkaline chlorides (?).

It is difficult, according to Riche, to admit that the hydrochloric acid is present in an absolutely free state, for calcium carbonate does not completely remove the acidity of the gastric juice; and if this be submitted to distillation, the hydrochloric acid comes over only as one of the final products. The fact has also been directly established that the albuminoid substances unite with mineral acids, forming compounds possessing an acid reaction, and which have lost certain properties of free acids.

R. Maly (1—173, 227) has come to the conclusion,

through experiments lately made as to the origin of the acid in gastric juice, that pure gastric juice contains no lactic acid, and that the origin of the hydrochloric acid of the stomach is not due to the decomposition of the chlorides present by lactic acid. The source of the free hydrochloric acid of the stomach is, according to Maly, to be sought in a disassociation of the chlorides, without the intervention of an acid.

Charles Richet, however (62—'77), has been studying the properties of the human gastric juice upon the person of the patient on whom Verneuil successfully performed gastrotomy. He has reached the following conclusions: 1. The acidity of the gastric juice, whether pure or mixed with food, is equivalent to 1·7 grammes of hydrochloric acid to a thousand grammes of fluid. 2. Acidity increases slightly at the end of digestion, and is independent of the quantity of liquid contained in the stomach. Wine and alcohol increase, but canesugar diminishes it. 3. If acid or alkaline matters are introduced, the gastric juice tends to return to its normal acidity. 4. The mean duration of digestion is from three to four and a half hours or more. Food does not pass successively, but in masses. 5. According to four analyses made by a modification of Schmidt's method, it was proved that free hydrochloric acid exists in the gastric juice. 6. It is possible to extract all the lactic acid contained in the stomach, and to prove that there is one part lactic acid to nine parts hydrochloric acid. 7. Following the method of Berthelot, that is, by agitation with anhydrous ether and deprived of alcohol, it can be shown that lactic acid is free in the gastric juice. 8. The question so long in controversy as to the nature of the free acid in the stomach seems, according to Richet, almost solved, and it may be said that in every 1,000 grammes of gastric juice there are 1·53 grammes of hydrochloric acid and 0·43 of lactic acid.

PEPSIN.—Observation has shown that if an infusion of the mucous membrane of the stomach be made, and the liquid acidified, it dissolves albuminoid substances as well as the gastric juice. The mucous membranes of the other organs do not possess this property.

Wasman was the first to extract from the mucous membrane of the stomach the active agent of this transformation.

It is an albuminoid substance known by the names of *pepsin*, *chymosin*, and *gasterase pepsin*. The best method of preparing pepsin is that of Schmidt. Gastric juice is neutralized with lime-water, filtered, evaporated to the consistency of syrup, and precipitated with concentrated alcohol. The pepsin is re-dissolved in water, precipitated with lead acetate; the precipitate is collected and decomposed by a current of hydrogen sulphide. The sulphur is separated by filtering, and the liquid containing pepsin is evaporated to dryness at a low temperature. The method by Bruecke we omit, as it appears to give a less pure product.

Pepsin is a yellowish gummy substance, soluble in water. Dissolved in 50,000—60,000 parts of acidulated water, it dissolves coagulated white of egg in six or eight hours. Alone it does not possess this dissolving power.

Heat does not coagulate pepsin, but destroys its digestive properties.

The union of pepsin and an acid is necessary for digestion. The acids of the stomach have been replaced by most of the other acids, with success. Artificial digestion may be very easily produced with pepsin and an appropriate amount of acid, or better, with gastric juice itself. If the proportion of acid is too large the action is stopped. The most suitable temperature is $38°$ to $40°$; the action is very slow at $50°$ and at $12°$; no action takes place at $100°$ or at $5°$.

Liquid albumen does not coagulate in the stomach, it is dissolved, and the solution is not coagulable by heat or acids. Coagulated albumen is converted into a soft, nacreous substance, then into a sort of pulp, which is gradually dissolved. According to some chemists, albumen is absorbed directly.

Fibrin is more energetically attacked by this secretion than ordinary albumen, which seems to prove that fibrin is neutral, while the alkali of the albumen, by saturating the acidity of the gastric juice, retards its action. Fibrin of the blood is attacked more rapidly than that of the muscles. At first it swells, then changes into a grey powder, which is finally dissolved. Gluten is dissolved very readily. Liquid casein coagulates, and afterwards dissolves in the same manner as coagulated albumen.

The name *peptones* is given to the products of the transformation of albuminoid substances by the action of gastric juice. J. Murk has lately observed the presence in saliva of a peptone ferment (60-'76-1800).

Peptones are soluble in water, coagulated by alcohol, lead acetate, mercury bichloride, and tannin. They differ from albumen, inasmuch as their solutions are not coagulable by heat; peptones obtained from osseous and gelatinous tissues are not precipitated by potassium ferrocyanide.

According to Meissner, the gastric juice produces with albumen, at first, a substance, *parapeptone*, identical with syntonin, which afterwards forms various other peptones.

Peptone (a) is precipitated by nitric acid, also by potassium ferrocyanide, acidified with acetic acid.

Peptone (b) is precipitated by the latter of these reagents.

Peptone (c) is precipitated by neither reagent.

Hoppe-Seyler and Gautier doubt the existence of these peptones.

The gastric juice does not dissolve all the nitrogenous substances of the food. A portion escapes its action, and is subsequently transformed in the intestines. It is generally admitted that gastric juice has no action on fatty substances.

Starch is not affected by the gastric juice, but it seems to be substantiated that the saliva continues its action on these substances in presence of the gastric juice. (Bruecke. Gautier. Besancz. Ed. of 1878. p. 831.)

In catarrh of the stomach the mucous secretion is very abundant, and various organic acids are formed.

LIVER. BILE.

Besides bile there have been extracted from the liver glucose, a substance analogous to starch, called *glycogene*, fatty substances, and various nitrogenous products, viz., leucin, tyrosin, and xanthanin.

The quantity of bile secreted is quite large. Guinea-pigs secrete in twenty-four hours a quantity of bile amounting often to several times the weight of their liver. In order to extract it from these or other animals the gall bladder is emptied, or the biliary ducts are ligatured, and an opening made in them.

The bile before entering the gall-bladder is odourless. On remaining in this vesicle it acquires a strong odour, a bitter taste, becomes concentrated, and forms a viscous greenish or brown liquid. Its density varies from 1.025 to 1.033.

Its normal reaction is slightly alkaline. It is coagulated by acids; the coagulum is formed of two acids, *taurocholic* and *cholic*, or *glycocholic* acids.

Human bile contains from 9 to 18 per cent. of solid substances; a less quantity is found in the bile of the ox and pig. More than half of this residue is formed of combinations of the acids just named, with different bases, though mainly with soda.

The bile may therefore be regarded as essentially a saponaceous compound. The other solid constituents

COMPOSITION OF BILE.

of the bile are—a neutral organic substance called *cholesterin*, a colouring matter, neutral fatty substances and salts; urea is sometimes found in it.

Strecker has extracted a base from bile which he calls *cholin*. This substance is identical with *neurin*, which has been extracted from the brain and yolk of egg. Its formula is $C_5H_{15}NO_2$.

COMPOSITION OF BILE.

(GORUP-BESANEZ.)

	Man of 49 years, decapitated.	Woman of 29 years, decapitated.	Man of 68 years, killed by a fall.	Child of 12 years, died of an injury.
Water	822.7	898.1	908.7	828.1
Fatty substances	177.3	101.9	91.3	171.9
Salts of the biliary acids	107.9	56.5	73.7	148.0
Fat	47.3	30.9		
Cholesterin				
Mucus, colouring matters	22.1	14.5	17.6	23.9
Mineral salts	10.8	6.3	—	—

The ash of ox gall contains:—

Sodium chloride	27.70
Sodium phosphate	16.00
Potassium ,,	7.50
Calcium ,,	3.02
Magnesium ,,	1.52
Ferric oxide	1.52
Silica	0.36

Small quantities of nitrogen have also been found,

and considerable proportions of carbon dioxide; this last gas may be extracted by a mercury pump. Animal food augments the quantity of carbon dioxide.

ACIDS OF THE BILE.—Human bile contains much more taurocholic than glycocholic acid.

The former alone exists in the bile of the dog; it abounds in the bile of serpents and fishes. Glycocholic acid is wanting in carnivorous animals. Both exist abundantly in the bile of the ox.

The bile of the pig contains special acids: hyoglycocholic acid, and taurohyocholalic acid.

In order to obtain the two acids of the bile, neutral lead acetate is added to ox-gall, which precipitates the glycocholic acid as a lead salt. This compound is collected, washed, boiled with 85 per cent. alcohol, and the boiling liquid filtered. It is then exposed to a current of hydrogen sulphide while yet warm; the lead sulphide is thrown on a filter and washed until the liquid becomes turbid. The glycocholic acid precipitates out of the solution, and is purified with boiling water.

The alkaline taurocholate is not precipitated by the lead acetate. To the first liquor lead subacetate is added until the precipitate takes on a fatty consistency; this precipitate is collected, washed, and suspended in water. A current of hydrogen sulphide is passed through the water, the liquid filtered and evaporated. The taurocholic acid is deposited as a white powder.

GLYCOCHOLIC ACID, $C_{26}H_{43}NO_6$, forms white needles moderately soluble in alcohol. One part is soluble in

100 parts of boiling and 300 parts of cold water. With alkalies and barium it forms soluble crystalline salts.

Boiling alkaline solutions and dilute acids, separate it into cholalic acid and glycocol by combining with water.

$$C_{26}H_{43}NO_6 + H_2O = C_{24}H_{40}O_5 + C_2H_5NO_2$$
Glycocholic acid.　　Water.　　Cholalic acid.　　Glycocol.

On being boiled with concentrated hydrochloric acid or sulphuric acid, it furnishes the following products:

Cholonic acid . . . $C_{26}H_{41}NO_5$
Choloïdic . . . . $C_{24}H_{38}O_4$
Dyslisin . . . . $C_{24}H_{36}O_3$

TAUROCHOLIC ACID, $C_{26}H_{45}NO_7S$, has not yet been obtained crystalline. It dissolves in alcohol and water, imparting to these an acid reaction. It is partly destroyed by the evaporation of its aqueous solution. It combines with one molecule of water on being boiled with alkaline solutions, cholalic acid and taurin being formed.

$$C_{26}H_{45}NO_7S + H_2O = C_{24}H_{40}O_5 + C_2H_7NO_3S$$
Taurocholic acid.　　　　Cholalic acid.　　Taurin.

THE BILE FERMENT.—W. Epstein and J. Müller

(60-1875-679) have lately investigated the influence of different substances upon the action of the ferment of the liver. Dilute aqueous solutions of carbolic acid (1 : 300) do not prevent the transformation of the glycogen into sugar if brought into contact with fresh, finely-chopped liver; yet this carbolic acid solution protects the liver from putrefaction for a long time. Five per cent. solutions of sodium chloride and sodium sulphate do not prevent or influence the transformation of the glycogen of the liver. Alkalies render the change slower, acids prevent it entirely; even when very dilute they greatly retard it. The action of acids, however, is only transitory; on neutralizing them the action of the ferment at once begins. Whether carbon dioxide prevents fermentation or not, has not been ascertained with certainty. The supposition of Tiegel that the change of the glycogen of the liver into sugar is connected with the destruction of the blood-corpuscles was not confirmed by the experiments of Epstein and Müller. They prepared from liver—moistening it with carbolic acid, drying at 30°, extracting with glycerine, and precipitating with alcohol—a ferment peculiar to liver, which converts glycogen into sugar very rapidly and easily.

TAURIN, $C_2H_7NO_3S$.—This substance may be prepared by boiling ox-gall with an excess of hydrochloric acid for several hours. Filter and add to the liquid five or six times its weight of boiling alcohol, and allow to cool slowly. The taurin, which is almost insoluble in alcohol, will separate out in colourless rhomboidal prisms.

Taurin has been produced artificially by Strecker, on heating isethionate of ammonia at 200°. This salt loses one molecule of water, and taurin remains.

$$\left.\begin{array}{l}H \\ (C_2H_4,SO_2)'' \\ NH_4\end{array}\right\} O_2 = H_2O + [(C_2H_4, SO_2)'', HO]' \left.\begin{array}{l} \\ H_2\end{array}\right\} N$$

This substance is, therefore, an amide like glycocol.
Taurin is found in the muscles of certain mollusks.

CHOLESTERIN. — $C_{26}H_{44}O, H_2O$. — This substance is widely diffused in the animal organism. Biliary calculi are almost entirely formed of it; it is found in the blood, yolk of eggs, spleen, pus, in various tumours, in the nerves and brain. It is easily extracted from biliary calculi, which are pulverized, suspended in alcohol with animal charcoal, and the mixture brought to boiling; after some time, the liquid is filtered. The cholesterin deposits on cooling.

Berthelot has shown that cholesterin has been wrongly classed among the neutral fatty substances; it is not saponifiable. He considers it a monatomic alcohol.

$$\left.\begin{array}{l}C_{26}H_{43} \\ H\end{array}\right\} O.$$

He has prepared the ethers of cholesterin by the action of acids:—

Acetic ether $\left.\begin{array}{c}C_{26}H_{43}\\C_2H_3O\end{array}\right\}O$.

This alcohol is dehydrated by anhydrous phosphoric acid, producing the carbo-hydride:—

$C_{26}H_{42}$ cholesterilene.

Cholesterilene is colourless, odourless, and tasteless, crystallizable in brilliant rhomboidal tablets, fusible at 145°, and volatile at 360°. Water does not dissolve it; it is slightly soluble in cold alcohol, and quite soluble in boiling alcohol and ether. It is soluble in the taurochlorates. It turns the plane of polarization to the left.

Heated with a few drops of nitric acid it becomes yellow, and this yellow substance, on being touched with a drop of ammonium hydrate, turns red. Sulphuric acid colours cholesterin red; if chloroform is added, a blood-red colour is obtained which, before disappearing, becomes successively violet, blue, and green.

According to Flint, cholesterin is an excrementitious substance, which results from the disintegration of nervous tissue, as it is not found in the blood entering the brain, but is found in the blood of the veins which leave it. Also, though absent in muscular tissue, it is always found in the nerves. It is absorbed by the blood and eliminated by the liver, as it is abundant in the blood of the hepatic artery and the vena porta,

COLOURING MATTERS OF THE BILE. 257

while little or none is found in the blood of the sub-hepatic veins. During digestion it is changed into a substance called *stercorin*, and evacuated in this state; also, when cholesterin is not discharged into the intestines, a decrease in the production of stercorin is observed. The retention of cholesterin in the blood gives rise to the serious malady cholesteremia.

COLOURING MATTERS OF THE BILE.—The bile furnishes two colouring matters: one brown, called *bilirubin*, *cholepyrrhin*, or *bilifulvin*; the other green, called *biliverdin*.

Bilirubin, $C_{16}H_{18}N_2O_3$.—This substance may be prepared by agitating fresh bile with water, ether, and dilute hydrochloric acid, which do not dissolve it, then with chloroform; the bilirubin dissolves, and is deposited on evaporation in orange-red crystals.

This body is dissolved by alkalies. It forms with lime a sort of lake, sometimes also found in the body (biliary pigment).

This substance has been found not only in the liver, but also in the brain, in cases of hæmorrhage, and in the placenta of dogs.

Biliverdin, $C_{16}H_{20}N_2O_5$, appears to be a product of alteration of bilirubin. It is first formed in the putrefaction of bile, is then changed spontaneously into biliprasin, $C_{16}H_{22}N_2O_6$ (?).

Biliverdin may be prepared by allowing an alkaline solution to stand for a time in the open air; this solution is then precipitated with hydrochloric acid. Both colouring matters are precipitated in this manner, and the biliverdin may be removed by treating the precipitate with alcohol, which dissolves this substance alone. Stoedler announces that he has extracted from bile two other colouring matters—bilifuscin and bilihumin. The former has been recently studied by Simony (111--73-181).

ACTION OF THE BILE ON FOOD. — The bile is not simply extracted from the blood by the liver, but is elaborated by it; the biliary acids are not found in any other part of the body, and the blood, in passing through the liver, loses its fibrin and a portion of its albumen. It has also been proved that the bile is formed in the liver, by removing this organ from frogs; when, after this operation, biliary acids were no longer found. Lehmann believes that the fibrin, wholly or in part, taken up by the liver, is transformed into glycogen.

The bile neutralizes the gastric juice, yet this saturation is not complete; the acids of the bile thus liberated have perhaps a certain utility in the intestines.

The bile has no digestive action on amylaceous matters, but assists in the digestion of fatty substances. Messrs. Schmidt and Bidder have shown that dogs assimilate, per kilogramme and per hour, under ordinary conditions, 0.465 gramme of fat, while only 0.093 gramme is absorbed when the bile has been removed through a fistula. The chyle of a dog fed with fat contains at least 3 per cent. of this matter; if the action of the bile be prevented by a fistula, this quantity will fall below 1 per cent.

On agitating bile with oil, it forms a rapid and

ACTION OF THE BILE ON FOOD. 259

persistent emulsion. Oils rise higher in a capillary tube when moistened with bile than when moistened with water.

Bile has no action on albuminoid substances in their ordinary condition. It precipitates acid solutions of albuminoid matter, but an excess of bile re-dissolves these precipitates. It is therefore not impossible that the bile takes part in the digestion of the albuminoid substances, acidified but not absorbed in the stomach.

Bile is found throughout the smaller intestines; it attaches itself to their folds, and by its adhesive character retains the non-absorbed food, and facilitates the action of the intestinal fluids.

Bile is not found in the large intestine, although we find there cholalic acid, taurin, and dislysin; glycocol has not been found. The excrements contain also taurin, dislysin, and cholalic acid, but Hoppe-Seyler, by determining the amount of this latter acid in the excrements, has shown that the greater part disappears in the intestine.

The bile appears to prevent putrefaction of the contents of the intestine.

The bile, therefore, after what we have stated, would seem to be a secretion, and also an excretion.

But little is known in regard to the formation of the immediate principles of the bile. We owe to Lehman an ingenious and probable hypothesis regarding the formation of the acids of the bile.

According to his theory the fatty substances, especially olein, play an important part in their production. In fact, the cholalic acid, like oleic acid in contact with alkalies, is broken up into an acetate and palmitate. Also the blood in passing through the liver loses fat; the amount of bile increases when the food is rich in fatty and nitrogenous substances, and the amount of fat increases as the bile diminishes and diminishes as the bile increases.

The bases to which these acids are united are derived from the blood, for it has been proved that the blood contains less salts on leaving the liver than on entering it.

The nitrogen of these acids, in the taurin and glycocol, is obtained from the albuminoid matters, as the blood, in passing through the liver, leaves behind a notable quantity of these substances. The sulphur of these products has the same origin.

Bilirubin appears to have great analogy with hæmatoïdin, which results from the alteration of the colouring matter of the blood; hence it would seem rational to admit that the colouring matter of the bile is derived from that of the blood, and that the hæmoglobin is destroyed in the liver. This explains why no blood is found in the bile.

The biliary secretion augments two or three hours after eating, and increases up to the thirteenth or fourteenth hour. Vegetable food produces bile in less quantity and less concentrated than animal food.

Fatty aliments, mixed with nitrogenous substances, increase both the amount of the bile and its richness in solids.

The injection of calomel increases the secretion of bile.

The biliary substances diminish in diabetes, in tuberculous affections, in dropsy, and typhus; increase in choleric persons, and in diseases of the heart and abdomen. The biliary secretion diminishes in fevers.

BILIARY CALCULI.—These are divided into biliary or cystic, hepatic, and hepato-cystic calculi, according to their origin.

They are composed of cholesterin, mixed with the colouring matters of bile and mucus. Cholesterin forms 80 per cent. of these calculi. To extract the cholesterin the powdered calculus is treated with boiling alcohol; on cooling, beautiful nacreous blades of cholesterin separate out.

Ox bile (gall) is employed for removing grease. It may be prevented from putrefying by evaporating to the consistency of syrup.

We shall speak of glycogen under the head of nutrition.

PANCREATIC JUICE.

The pancreatic juice is a liquid, colourless and somewhat viscous, having a saline taste. Its density is not uniform, as it contains variable proportions of solid matter, which have been found to amount to as high as 11 per cent.

Its reaction is alkaline, and due to sodium hydrate. The most important proximate principle of this juice is an albuminoid substance called *pancreatin*. In it is likewise found a fatty substance, also leucin, tyrosin, xanthin, and several salts, among which are sulphates and chlorides.

This juice owes to the pancreatin present its property of coagulating with heat, alcohol, and acids. This fact led to the belief, formerly, that the albuminoid principle of this juice was albumen; this is not true, however, as the coagulum formed by alcohol re-dissolves in water and re-assumes the viscous appearance and the characteristics of pancreatic juice.

Pancreatin is prepared by pouring 85 per cent. alcohol into pancreatic juice. White flakes are formed, which are soluble in water, yielding a solution which possesses, to a high degree, the property of converting starch into sugar. Jenneret states (18–77–389) that the action does not require oxygen.

Pancreatic concretions contain variable proportions of nitrogenous organic matter and calcium carbonate and phosphate.

ACTION OF PANCREATIC JUICE.—This juice appears to act upon the three classes of organic aliments; it promptly forms an emulsion with neutral, fatty substances, and is even capable of saponifying them. Its action is most rapid at about $35°$; its action is arrested by acids, even the acidity of the gastric juice retarding its action. It has been found, also, that in chyle neutral fatty substances predominate over acid fats, and it is therefore believed that the pancreatic juice renders fats assimilable by forming an emulsion with

ACTION OF PANCREATIC JUICE.

them. The bile and intestinal secretion share with the pancreatic juice this property, for it has been shown that the chyle contains emulsions of fat, after the ligature of the pancreatic duct : their action, however, is very weak, as Bernard found that if the pancreatic juice be prevented from entering the intestines, the greater part of the fatty substances are found unchanged in the excrements.

Corvisart, Kühne, and others have shown that this juice dissolves fibrin and coagulates albumen, transforming them into assimilable products, analogous to the peptones. It, however, will act alone, whatever may be the state of the liquid, while pepsine requires the co-operation of an acid. According to Schiff the functions of the spleen are connected with those of the pancreas, as the pancreatic juice has no action on albuminoid substances after the spleen has been removed.

Moreover, and this, according to Bouchardat and Sandras, is its principal *rôle*, the pancreatic juice is the chief agent in effecting the transformation of farinaceous food.

The transformation of starch into glucose is slow, as farinaceous matter has been found in the intestines twenty-four hours after eating. It is probable that only a small quantity of the starch is absorbed in the form of glucose, the greater part being normally absorbed in the form of dextrine. The transformation continues, absorption having been accomplished, under the action of the ferment absorbed at the same time, with the dextrine.

Pancreatic juice is rapidly decomposed in contact with the air.

Claude Bernard states that an infusion of pancreas, or a solution of pancreatic juice, after having stood in the air for some time, assumes an intense red colour on the addition of chlorine water. Nencki (18–'78–79) is of the opinion that pancreatic digestion is essentially a process of putrefaction.

INTESTINAL FLUIDS.

These liquids are complex products even when the ducts conducting the bile and pancreatic juice to the intestines are closed, as there are several varieties of glandular apparatus which secrete liquids throughout the length of the intestinal canal. Colin has shown that mucus is also secreted. This physiologist, by binding the intestine at two points, about two metres apart, was enabled to obtain about one hundred grammes of the liquid secreted by the glands of Lieberkühn, and having removed the mucus by deposition and filtration, he examined its properties.

It is a limpid liquid, slightly yellowish, secretion, whose density is 1.010 and reaction very alkaline. Saturated with an acid it is coagulated by heat. It is also coagulated by alcohol and precipitated by lead acetate.

This fluid continues the transformation of farinaceous substances into dextrine and sugar, and forms emulsions with fatty matters. Although possessing an alkaline reaction, it acts upon albuminoid substances. Thus, according to Bidder and Schmidt, flesh and albumen coagulated by heat, and enclosed in the intestines by ligature, soften, dissolve, and are digested; consequently, the intestinal fluid completes the digestion of nitrogenous substances: it is not known what constitutes its active principle. Thiry found in pure intestinal fluid from a dog:

Water	97.585
Albuminates	0.802
Other organic substances	0.734
Inorganic substances	0.879

The gases of the smaller intestines are chiefly carbon dioxide and hydrogen. In the larger intestine these gases are mingled with methane, and traces of hydrogen sulphide; the methane amounts to as high as 50 per cent. of this volume when the food is vegetable.

The excrements contain 10 to 15 per cent. of solid substance, of which 6 to 7 per cent are mineral. In them has been found *stercorin* or serolin, which is a fatty non-saponifiable matter, a product of the decomposition of cholesterin, also a white crystalline substance, called *excretin*, which contains sulphur, and which probably effects the elimination of this element from the system.

Calcareous and magnesian phosphates, sodium

chloride, a small quantity of silica, fatty matter, products of the decomposition of the acids of the bile, of the epithelium, and the tissues of the vegetables are also found.

The use of iron preparations colours the excrements blackish-green (iron sulphide). Calomel gives them a light green colour. If they contain blood the colour will be dark or nearly black.

Cholera excrements contain coagulated albumen, cystoid corpuscles, and chlorides; common salt amounting sometimes to over one-half the total weight. In *dysentery* and in Bright's disease mucus is found. In certain excrements the presence of alloxan, a product of the oxidation of urea, has been detected. In *typhoid fever* they are mostly fluid and alkaline. On standing a viscous mass deposits, containing mucus, food *débris*, and generally crystals of magnesio-ammonium phosphate. The fluid above the deposit contains albumen, various soluble salts and biliary constituents. Addition of nitric acid produces a rose-red coloration, as is also the case in cholera stools.

Subjoined are the results of two analyses of human excrements, which from the inherent difficulties of such investigations cannot be regarded as exhibiting their composition with very complete accuracy. The one by Wehsarg is of recent date:—

		Berzelius.	Wehsarg.
Water		75.30	73.300
Biliary salts	0.90		
Mucus and biliary resins	14.00		
Albumen	0.90		
Extractive matters	5.70		
Aqueous extract		5.340	
Alcoholic „		4.165	
Etherous „		3.070	
Food debris	7.00	8.300	
Mineral salts	1.20		
Earthy phosphates		1.095	
Total salts		29.70	21.970

INTESTINAL CONCRETIONS.—These contain a large proportion of fatty matter, a substance analogous with fibrin, calcium phosphate, and sodium chloride.

The name *bezoar* is given to the intestinal concretions found in gazelles and goats. They are formed sometimes of an organic (lithiofellic acid, sometimes of calcium and ammonio-magnesium phosphates.

SUMMARY OF DIGESTION.

To recapitulate, the food is mechanically divided in the mouth by the action of the tongue, teeth, and the saliva, which latter commences the transformation of the starchy matter. The bolus formed passes through the œsophagus, and arrives in the stomach, where the digestion of the greater part of the nitrogenous substances is effected by the action of the gastric juice. The majority of these substances having become assimilable, are absorbed by the walls of the stomach, and the remainder of the food passes into the duodenum. There the emulsion of the fatty matter is prepared, and the transformation of the starch into glucose effected by the action of the bile, pancreatic juice, etc. This latter fluid also effects the digestion of the nitrogenous matters.

The food as modified by these different changes forms *chyme*. It now enters the jejunum, and moves forward by peristaltic and muscular motions. It here receives the secretions, which complete the transformation of starch into sugar, the solution of the albuminoid matter, and the emulsion of the fats. The chyliferous vessels absorb almost exclusively these latter substances, while the intestinal veins absorb the products of the transformation of the fluids and albuminoid bodies.

The absorption of the liquid products of digestion, and, in general, the absorption of liquids, is effected by means of a very complex mechanism.

Diffusion takes the principal part in this process; in fact the animal membranes are lined with colloid cells, through which diffusion takes place with great rapidity, and we have seen that, although albuminoid substances are but slightly dialyzable in a natural state, they become quite readily so on being transformed into peptones.

ABSORPTION.

CHYLE, LYMPH.

A very considerable quantity of lymph and chyle is constantly poured into the blood. These fluids are very analogous in character: they have a circulatory movement; they are formed of a liquid (*serum*) in which float globules capable of uniting to form a *clot* or *coagulum*; their composition is also similar, there being, in fact, little difference, except in the proportion of their elements.

The chyle is a lactescent fluid contained in special lymphatic vessels, into which it passes directly from the intestines; it accumulates in the mesenteric glands, whence it passes into the thoracic duct. It may be obtained by opening this duct and ligating the same near where it enters the sub-clavian vein. However, at this point it has already undergone elaboration, and is mingled with lymph coming from different points in the body. Before describing the chyle, it should be remarked that the knowledge we possess of this fluid is based chiefly upon investigations among the lower animals.

The chyle of an animal deprived of food is yellowish; during digestion, especially of fatty food, it is lactescent. This appearance is due to the fatty bodies present, for if it be agitated with ether it loses its milky appearance. It has a feeble odour, a slight taste, and its reaction is faintly alkaline.

Chyle contains fibrin, albumen, and urea. The presence of casein is suspected, but not certain; yet the albumen of the chyle is more alkaline than ordinary albumen. The serum of chyle becomes covered with a film during evaporation; it coagulates only in small flakes, and acetic acid precipitates it but partially. Chyle separated from the body coagulates in ten to fifteen minutes, producing a clot floating in an albuminous liquid; this coagulation is due to the fibrin present.

Lymph is a colourless, or nearly colourless, liquid. It reaction is alkaline, which appears due, like the alkalinity of chyle, to a matter analogous with casein. It contains white globules, fibrin, albumen, urea, fatty bodies, and salts, which are chiefly lactates.

It coagulates after being exposed to the air for a few minutes, producing a thin, soft clot, coloured red by globules of blood.

Robin found the composition of human lymph and chyle to be, in 1000 parts, as follows:—

	Lymph.	Chyle.
Water	960 to 965	900 to 990
Sodium chloride	4 ,, 6	5 ,, 7
Sodium carbonate	1 ,, 2	not determined
Calcium carbonate Alkaline and calcareous phosphates	0.05 ,, 2	0.80 to 3
Alkaline sulphates	0.23 ,, 0.50	not determined
Crystallized organic principles (urea, glucose)	3 ,, 8	5 to 9
Fatty bodies	2 ,, 9	10 ,, 36
Albumen	33 ,, 60	30 ,, 40
Fibrin	1 ,, 5	3 ,, 4
Peptone	3 ,, 4.5	6 ,, 8
Hematosin	0·6	0.6

Wurtz has found urea in the lymph of various animals.

The above analyses do not wholly agree with those of other chemists, and from the variable character of these two fluids, and the inherent difficulty of their analysis, the foregoing figures must be considered as giving only an approximative idea of their chemical composition. The variations in composition are greater, however, in chyle than in lymph.

RESPIRATION.

THE BLOOD.

The blood is at once the nutritive and the purifying fluid of the body. From one part of the body it gathers the liquids elaborated by digestion, and in another it takes from the air its vital principle, oxygen, to act upon these liquids; also it collects in different parts of the body the various effete products, and carries them to the organs destined to eliminate them. The blood also serves to distribute heat throughout the body.

It circulates incessantly in the capillaries, arteries, and veins. Arterial blood is vermilion red; venous blood is reddish brown. Its odour varies somewhat with the species, and seems more marked in the male than in the female. According to Barruel, sulphuric acid increases its odour.

Its taste is slightly saline. Its density varies between 1.045 and 1.075. It has an alkaline reaction, which is due to sodium compounds.

On placing under the microscope a very thin membrane like the foot of a frog, it may be seen that the blood has a rapid circulatory movement, and that it is formed of a colourless liquid (*plasma*), in which floats an immense number of globules, drawn with it in the circulating current. The globules are microscopic in size, the majority are red, yet there are some which are colourless.

A great many analyses of blood have been made. The results vary according to the physiological conditions of the subject, but the following tables give an average result:—

	Venous Blood.	
	Man.	Woman.
Water	780.00	791.00
Globules	140.00	127.00
Albumen	69.00	70.00
Fibrin	2.20	2.20
Extractive matter and salts	6.80	7.40
Serolin	0.02	0.02
Fatty matters containing phosphorus	0.49	0.46
Cholesterin	0.09	0.07
Salts of fatty acids	1.00	1.05
Loss	.40	.80
	1000.00	1000.00

Salts contained in 1000 grammes of blood:—

Sodium chloride	3.10	3.90
Other soluble salts	2.50	2.90
Iron	0.565	0.541
Phosphates	0.330	0.354

(Becquerel and Rodier.)

The blood on leaving the body loses its fluidity in a few minutes, becomes viscous, and changes into a gelatinous mass which gradually contracts and forces out drops of liquid, *serum*, which unite around the *coagulum* or clot. This clot gains in consistency, and after ten to thirty hours it ceases to contract.

	Composition.
Serum	870
Clot	130
	1000

Each of the two parts composing the blood has the following composition:—

Clot	Fibrin 3	} 130
	Globules. 127	
Serum	Water	790
	Albumen	70
	Oxygen	
	Nitrogen	
	Carbon dioxide . .	
	Extractive matter . . .	
	Phosphorated fat . . .	
	Cholesterin	
	Serolin	
	Margaric acid	
	Sodium chloride . . .	
	Potassium chloride . . .	
	Ammonium chloride . .	10
	Sodium carbonate . . .	
	Calcium carbonate . . .	
	Magnesium carbonate . .	
	Calcium phosphate . . .	
	Sodium phosphate . . .	
	Magnesium phosphate . .	
	Potassium phosphate . .	
	Sodium lactate . . .	
	Salts of fixed fatty acids . .	
	Salts of volatile fatty acids .	
	Yellow colouring matter. .	
		1000

(Dumas).

Many other substances also exist in the blood. We may say, in a general way, that it contains most of the immediate principles which compose the tissues and liquids of the body.

COAGULATION OF THE BLOOD—SERUM.—The blood, we have said, coagulates on being removed from the body. This coagulation seems to be due to the fibrin, for if the blood be beaten with twigs, the fibrin is seen to attach itself to the branches, and the blood has lost its property of coagulating. The serum of the coagulum is not therefore identical with the plasma.

This latter contains fibrin, and the former has been freed of it. The fibrin imprisons the globules of the blood, and these together form the coagulum.

Coagulation is not due to the fact that the blood remains at rest on leaving the body of the animal, or because it becomes cooled, for by keeping the blood in motion and maintaining the temperature of the body, solidification is not arrested. It is not due to the presence of air, as coagulation takes place in other gases and in a vacuum. Acids coagulate blood. The rapidity of coagulation varies from a few minutes to several hours. It is slower in the blood of the vigorous than in that of weak persons. It is accelerated by raising the temperature from 30° to 48°.

It is retarded several hours by lowering the temperature to 0°. The addition of albumen, sugar, and solution of alkaline salts produces the same effect, and coagulation is even arrested by concentrated solutions of certain salts, especially sodium sulphate.

If pulverized sodium chloride be added to this liquid

it furnishes flakes of an albuminoid substance, which, according to Dennis, is different from albumen and fibrin. He has given it the name of *plasmin*.

It is very soluble in water, and is easily decomposed into soluble and insoluble fibrin, which, according to this chemist, is the cause of coagulation.

Virchow and Schmidt regard fibrin as produced by the combination of two albuminoid principles of the blood, the *fibrino-plastic* substance or *paraglobuline* and *fibrinogen* or *metaglobuline*, at the moment when the blood is removed from the body.

These two bodies may be obtained by passing a current of carbon dioxide through plasma, diluted with ten times its volume of water at 0°. The fibrinoplastic substance is immediately precipitated in white flakes, which are collected on a filter and washed with water, charged with carbon dioxide, as aerated water dissolves it. The stream of carbon dioxide is now allowed to pass through the liquor for a long time. At first an abundant foam is formed, then the fibrogene separates out as a glutinous mass.

If these two substances are separately dissolved in water which is slightly alkaline, and are then mixed, a gelatinous matter separates out which soon forms into filaments, analogous in appearance to fibrin.

According to Schmidt these two substances require for their action the presence of a ferment, which is not developed in blood during circulation, but which is produced as soon as the blood is removed from the body; this ferment has not been isolated—whence is it derived?

According to others, the fibrin is already formed and solid in the blood, and coagulation is simply the result of the aggregation of these solid particles. Supposing it to be proved that the fibrin exists in a solid state in the blood, it yet remains to determine the cause of this aggregation in air.

It has been said that the fibrin surrounds or exists in the globules; since, however, we can separate the globules and still have a coagulable plasma, this hypothesis is not admissible.

Smée considers fibrin as oxidized albumen. But how can it be supposed that this oxidation takes place in a few seconds?

Notwithstanding all that has been written concerning the probable cause of the coagulation of the blood, it must be confessed that the causes thus far assigned are not wholly satisfactory. They are, for the most part, mere hypotheses.

Serum is chiefly a solution of albumen. But this albumen is found in different states, *free* and *combined with soda;* also, in the analyses above cited, the albuminoid substances (fibrogene and fibrino-plastic substance,) which are precipitated by carbon dioxide, have been considered as albumen.

E. Mathieu and V. Urbain (9-79, 665 and 698) seem to have established, though disputed by A. Gautier (9-83-277), that the coagulation of blood is caused by the carbon dioxide, which, when blood is exposed to the air, is expelled from the blood globules, in which it is contained during life, by the oxygen of the air. Hence it is clear why alkalies and ammonium hydrate, as well as concentrated solutions of certain salts which absorb carbon dioxide, prevent the coagulation of blood.

Venous serum contains somewhat more water than that of the arteries, the serum of women containing, according to C. Schmidt, more water than that of men. The proportion of water increases in most diseases; the reverse is seldom observed except in certain fevers and in cholera.

The abundance of albumen in the serum and in the blood in general proves that this substance is the principal constituent of the albuminoid fluids and nitrogenous tissues of the body. Its proportion ranges between 63 and 70 in 1000 parts; it increases at the moment of digestion. Venous blood contains more albumen than arterial blood. Its quantity generally diminishes in disease; yet it increases, as does the fibrin from other causes, in inflammatory fevers.

The fatty bodies of the serum are often crystallizable, and it was a mixture of these substances which was formerly called *seroline*. There is a small quantity of olein and oleic acid in the serum. There is also found in it stearin, margarin, the two corresponding acids, and cholesterin. The venous blood contains more of this last body than the arterial blood; the blood of the vena porta contains more than that of any other vessel.

The amount of fatty bodies increases during digestion. They diminish in general during disease, with the exception of cholesterin, which often increases. The blood of women contains a little more than that of men.

Glucose always exists in the serum; its proportion is very small; it increases during digestion if the food is very starchy. The blood of the hepatic veins contains a considerable proportion of this substance, while the blood of the vena porta hardly contains any whatever. The blood of diabetic persons scarcely furnishes 0.05 per cent; normal blood contains at the most 0.0020 per cent.

The blood which is most rich in salts is that of the vena porta; arterial blood in general contains more than venous blood.

A considerable diminution in the quantity of sodium chloride in food affects health seriously.

Many other substances have been found in the serum. Some are constantly met with; these are urea, uric acid, hippuric acid, creatin, creatinin, casein, acetic acid, dextrin, and glucose, the peptones, sodium and potassium chlorides, sodium carbonate and phosphate, sodium and potassium sulphates. Neither glycocol, leucin, taurin, nor tyrosin has been found.

Prevost and Dumas detected the presence of urea in the blood after the suppression of the urinary secretion. The existence of this body in the blood has been proved by Béchamp and other experimenters. According to Picard, normal blood contains 0.017 of urea; twice as much is found in the renal artery as in the renal vein.

Casein exists principally in the blood of pregnant women, nurses, and nurslings.

In leucocythæmia the blood contains gelatin, hypoxanthin, lactic and formic acids: biliary acids in diseases of the liver, ammonium carbonate in persons having cholera.

THE COAGULUM—*crassamentum* or *clot*—is red and somewhat elastic. It is formed principally of fibrin and globules, and incloses about one-fifth of its volume of serum. It seems to form more rapidly in the blood of a child than in that of an adult, in that of women sooner than in that of men; its compactness is in inverse proportion to the rapidity of its formation.

In some pathological states the separation of the clot and serum does not take place, and a gelatinous mass remains. In others the blood is rich in fibrin, and a whitish matter called "buff," or buffy coat, is observed on the surface, which is fibrin nearly free from globules.

On agitating coagulum in a bag placed in a stream of water the globules and other proximate principles, with the exception of the fibrin, are carried away by the water, and the latter remains in the cloth in the form of greyish filaments.

GLOBULES.—Blood globules may be obtained by receiving fresh blood in a saturated solution of sodium sulphate, then filtering; the globules remain on the filter mingled with the solution of the salts.

The red globules of the blood of mammalia are

minute circular disks, slightly thickened at the margin. It is generally admitted that they are formed of a colourless membrane; they would, therefore, be veritable cells. Yet some observers regard them as an agglomerated gelatinous substance destitute of exterior membrane. This latter view is not probable, for on placing a drop of blood on the slide of a microscope and adding a little water the globules are seen to swell, also the margins become yellow in contact with a solution of iodine.

According to Béchamp and Estor, there exists in the blood on leaving the body an immense number of movable granulations of extreme minuteness, capable of development, of uniting and even of changing into bacteria and bacterides. These microscopic beings—called microcosms—are said to form the globules by their aggregation.(?)

These savants affirm that they have seen them form new cells, and that the blood-globules in the body are the result of the activity of the microcosms. The blood-globules in fishes, reptiles, and birds have an elliptic form.

Milne-Edwards has shown that no connection exists between the size of animals and the size of their blood-globules, but that they are smaller as the organism is more perfect and respiration more active.

Globules have a greater density than serum. Placed in contact with water they absorb the same, swell, and become spherical. At the same time a quantity of the colouring liquid of the globules is extravasated and colours the water. Change of form exerts a great influence upon their colour. On swelling, they take on a darker tint. On losing water they become clear and red; this takes place when they come in contact with sugar and alkaline liquids. The globules cannot, therefore, be collected on a filter and washed with water without becoming altered. A solution of sodium sulphate of 18° Baumé does not attack them, and if a mixture of one volume of blood and two volumes of this solution be thrown upon a filter, they may be separated from the serum without being destroyed.

This result is better obtained by adding to defibrinated blood ten times its volume of a concentrated solution of common salt; the globules are precipitated, and may be washed with salt water.

Besides red globules there exist in the blood white corpuscles; their number is much smaller (about 1 in 400). There appear to be two kinds.

The most abundant, the *plasmic*, *lymphatic*, and *fibrinous globules*, have a spherical form. Their border is very well defined; they contain a viscid matter in which float little nuclei, which refract light strongly. They are larger than the red globules (diameter $= 0.0113$ millimetre), also lighter than these latter.

They may be distinguished from the coloured globules by their different reactions. Water distends without destroying them, and dissolves them only after a long time. Acetic acid simply causes them to contract.

These globules are not, like the preceding ones, specially characteristic of the blood, for they are found in most of the other fluids of the system.

The name *globulines* has been given to certain white corpuscles, not numerous, whose diameter is about $\frac{1}{100}$ of a millimetre. They are small spherical nuclei, which are probably derived from the chyle.

The number of globules in a cubic millimetre of blood has been estimated at four to five millions.

ANALYSIS OF DRIED GLOBULES.

	Human blood.	Blood of a dog.
Hæmoglobin	86.79	86.50
Albuminoid matter	12.24	12.55
Lecithin	0.72	0.59
Cholesterin	.25	0.36

(Hoppe-Seyler).

The albuminoid matters appear to be constituted chiefly, if not wholly, of fibrino-plastic substance.

Red globules treated with water become spherical and distended, the colouring matter and other elements pass into the water, and there remains a gelatinous mass of a pale tint called *stroma*, which is formed chiefly of albuminoid substances.

HÆMOGLOBIN. — This substance is prepared by mixing defibrinated blood with an equal volume of water, and adding to this liquid one-fourth its volume of 80 per cent. alcohol ; this mixture is allowed to stand twenty-four hours exposed to a temperature of 0°.

Crystals then form in the liquid, which are pressed out on a filter and purified by re-dissolving in water and re-precipitating by adding to the solution one-fourth its volume of alcohol and exposing to a temperature below 0°.

It may be easily obtained in an impure state by adding ether, drop by drop, to defibrinated blood. The colour of the blood darkens on account of the destruction of the globules, and the liquid deposits crystals on exposure to a low temperature. This substance is also known as *hæmatocrystallin*.

The hæmoglobin of human blood forms regular rectangular prisms ; the same is true of that of the dog, cat, horse, and lion. That of guinea-pigs and mice crystallizes in tetrahedrons, and that of squirrels in hexagons.

It is insoluble in absolute alcohol, ether, chloroform, carbon bisulphide, and essential oils. Acids decompose it without dissolving. Alkalies dissolve it by altering its nature. It has a slightly acid reaction. It may be preserved after having been dried at a low temperature. In aqueous solutions it is slowly destroyed at ordinary temperatures, and instantly at 100°. It absorbs oxygen at ordinary temperatures, one gramme of hæmoglobin dried at 0° absorbing more than 1 c.c. In a vacuum nearly the whole of this gas again escapes. Hæmoglobin may therefore be considered as that constituent of the globules which fixes oxygen.

Hæmoglobin contains, besides carbon, hydrogen, oxygen, and nitrogen, small quantities of sulphur and phosphorus, and about 0.5 per cent. of iron.

HÆMATIN—HÆMIN.—An aqueous solution of hæmoglobin heated to about 75° or 80° is decomposed into another colouring matter, hæmatin, and an albuminoid matter which coagulates. This decomposition takes place gradually at ordinary temperatures, in presence of acids or alkalies in solution. Hæmatin represents only about four per cent. by weight of hæmoglobin.

If a small quantity of sodium chloride and strong acetic acid is added to hæmoglobin or blood, and after having heated this mixture over a water-bath, it is allowed to slowly cool, hydrochlorate of hæmatin (hæmin) is precipitated in rhomboidal crystals of a brown colour; this is also a characteristic test in medico-legal investigations. Virchow, also Robin, have designated as *hæmatoidin* a crystalline matter, containing neither iron, sulphur, nor phosphorus, and which results from the destruction of hæmatin in sanguinary effusions. This body is, however, now generally recognized as bilirubin.

Hæmoglobin forms with carbonic oxide a crystalline compound, which may be prepared in the same manner as hæmoglobin, by employing blood previously agitated with carbon oxide. These crystals have the same form as the hæmoglobulin.

F. Hoppe-Seyler (60-1874-1065) has lately carefully investigated the colouring matter prepared from hæmatin, by reducing substances, and proved its identity with the *urobilin* of Jaffé (36-1869-815), and the *hydrobilirubin* of Maly (36-1872-836).

It should be observed that Thudichum and Kingzett have quite recently (32-'76-255) made an analysis of hæmin, and finding the same to contain 7.65 per cent. iron, 3.02 chlorine, and 0.60 phosphorus, have come to the conclusion that hæmin is in reality a substance consisting of hæmatin, chlorhydrate of hæmatin, and a crystalline compound containing phosphorus, which they regard as identical with *myelin*, a body claimed by Virchow as existing in the brain.

C. Husson (9-81-477) states that crystalline compounds may be formed between hæmatin and phenol, oxalic acid, valerianic acid, tartaric acid, citric acid, and silica.

Hæmoglobin forms crystalline compounds with nitrogen dioxide and cyanhydric acid.

Red globules are not attacked by albumen, gum, or sugar solutions, carbon dioxide, or neutral salts of the alkaline metals. Alum, chlorine, sulphuric acid, and nitric acid cause them to contract; water, organic, and phosphoric acids, and alkaline solutions dissolve them.

Milne-Edwards (9-79-1268) remarks that the respiratory power of the blood depends upon the number of red blood-corpuscles present.

IRON IN THE BLOOD.—Boussingault determined this metal among the elements of cow's blood. In 100 parts he obtained:

	Total Mineral Substances.	Iron.
Dry fibrin	2.151 grammes	0.0466 grammes
Dry albumen	8.715 ,,	0.0863 ,,
Dry globules	1.325 ,,	0.3500 ,,

The colouring matter of the blood owes its colour mainly to the large proportion of iron in the globules, which, dried, gives:

10.750 per cent. ash, containing
9.043 ,, ferric oxide,
1.707 ,, other mineral substances,

formed almost entirely of lime and phosphoric acid.

P. Picard (9–79–1266) found the proportion of iron in the blood of dogs to be quite variable and proportional to the amount of oxygen the blood was capable of absorbing. In his investigations regarding the amount of iron in the human body, the spleen gave higher proportions than any other organ.

Jolly has very lately (61-'78) made analyses that appear to show that the iron in blood exists as ferrous phosphate.

GASES OF THE BLOOD.

Magnus was the first to make, in 1837, an extended study of the gases contained in the blood. A flask containing blood was agitated violently, in order to coagulate the fibrin. The defibrinated blood was transferred

GASES OF THE BLOOD.

into a bell glass, filled with mercury. He obtained the following composition of the gases liberated:—

	Venous Blood.	Arterial Blood.
Carbon dioxide	71.6	62.3
Oxygen	15.3	23.2
Nitrogen	13.1	14.5
	100.0	100.0

His methods of collecting the mixed gases were not, however, complete, and later analyses may be regarded as more reliable.

C. Bernard determined the amount of oxygen in the blood, profiting from a fact discovered by him that carbon oxide displaces the oxygen. The blood is taken directly from the body by a syringe, and immediately introduced into a graduated tube half-filled with carbon oxide. This is agitated, and kept at a temperature of 40°, after which the amount of oxygen in the gas is determined.

Venous and arterial blood dissolve variable quantities of oxygen. 100 volumes of blood from a young dog contained:

In the left ventricle, 23.0 vol. of oxygen. Animal fasting.

In the left ventricle, 17.6 vol. of oxygen. Animal digesting.

In the right ventricle, 10.0 vol. of oxygen. Animal fasting.

In the right ventricle, 10.2 vol. of oxygen. Animal digesting.

The gases from the blood of a dog gave, in 100 parts:

	Nitrogen.	Oxygen.	Carbon dioxide.	Carbon di-oxide combined.
Arterial blood	{1.61 {2.30	20.05 22.2	34.8 35.3	traces. 0.88
Venous blood	{1.32 {1.64	12.1 11.6	43.5 42.8	4.40 5.30

When venous blood is agitated with oxygen it takes on the red colour of arterial blood.? If, on the contrary, arterial blood be agitated with carbon dioxide, hydrogen, or nitrogen, it assumes the dark brown tint of venous blood.

P. Bert (9-80-733) found, in his investigations upon the power of blood to absorb oxygen at different pressures, that a compound of hæmoglobin with oxygen (oxyhæmoglobin,) is obtained when blood is agitated with air at ordinary pressure. Increase of pressure increases the proportion of oxygen in this compound: it also remains constant until the pressure is lowered to one-eighth of an atmosphere at 16°, but at the temperature of the bodies of mammalia it decomposes as the pressure is further removed.

The blood on leaving the lungs does not contain as much oxygen as it is capable of absorbing. Grehant has found that in agitating blood with oxygen the quantity which it is capable of absorbing is to the quantity ordinarily found in it as about 26 to 16. But there is a great difference in this regard between individuals, their state of health, etc.

The opinion has been expressed that the blood contains ozone, but this cannot be admitted, as the blood, like all organic matter, destroys ozone. It is only necessary to agitate blood in a vessel with ozone to obtain proof that these two bodies are incompatible, for the odour of ozone disappears immediately.

J. Dogiel (75–24–431) states, as the result of his recent researches regarding the action of ozone upon the blood, that the action of the ozone is chiefly upon the red blood-corpuscles; their colouring matter is expelled, and they become darker within fifteen minutes. After this change alcohol, ether, or chloroform produces no separation of crystals of hæmoglobin. Upon passing ozone through defibrinated blood for a long time, flakes separate out, which, after washing with water, are not to be distinguished from fibrin. By continued action of ozone blood becomes first of a dirty, yellowish-green colour, and, finally, colourless. Hæmatin is likewise rendered colourless by ozone. Blood poisoned with carbon oxide attains in a short time the properties of normal blood on exposure to the action of ozone, carbon dioxide being given off. Blood containing carbon oxide is discoloured less quickly than normal blood, and does not so quickly lose its property of depositing crystals of hæmoglobin. The change of the blood corpuscles produced by ozone should not be confounded with the change produced by carbon dioxide.

Carbon oxide displaces the oxygen of the blood, and is very deleterious when inhaled.

Chlorine coagulates blood, removes the iron which enters into the composition of its colouring matter, and subsequently destroys the organic matter. The iron is changed into ferric chloride, capable of being detected with reagents.

Arsenide of hydrogen completely changes the nature of blood, which assumes the colour of ochre.

Defibrinated blood becomes brown and then dark green under the action of hydrogen sulphide; the colouring matter is attacked and the globules destroyed.

Certain neutral salts, the alkaline sulphates, phosphates, and nitrates, redden the blood in the same manner as oxygen.

Oré (9-81-833, 990) asserts that acetic acid, sulphuric acid, nitric acid, hydrochloric acid, phosphoric acid, or alcohol after being diluted with water, may be injected into the blood-vessels of a living animal without producing coagulation of the blood.

DIFFERENCES BETWEEN ARTERIAL AND VENOUS BLOOD.

We have incidentally noticed these differences in studying the various constituents of the blood.

Longet sums them up as follows:

Arterial Blood.	Venous Blood.
1st. Vermilion red.	1st. Brown red.
2nd. Rich in fibrin.	2nd. Rich in albumen. ?
3rd. ,, ,, globules. ?	3rd. Has less water.
4th. ,, ,, salts.	4th. ,, ,, extractive matters.
5th. Contains about 30 parts of oxygen to 100 of carbon dioxide.	5th. Contains about 22 parts of oxygen to 100 of carbon dioxide.
6th. More coagulable.	6th. Less coagulable.
7th. Less abundant in fatty matters. ?	7th. Globules more abundant in fatty matter. ?
8th. Has the same composition in all parts of the arterial system.	8th. Has a different composition in different parts of the venal system.

We have indicated by ? such items in Longet's tabulation as are doubtful, or at least are not constant.

INDUSTRIAL USES OF BLOOD.—Coagulated blood serves as food in certain countries, as Germany, Sweden, and Italy. Freshly drawn blood is highly nutritious, and not unfrequently used by emaciated and greatly enfeebled invalids. The large quantity of albumen contained in the blood and the property which albumen possesses of coagulating on heating, causes blood to be employed in sugar refineries for the clarification of sugars.

CHEMICAL PATHOLOGY OF THE BLOOD.

Since the blood circulates throughout the entire body, it is evident that diseases which manifest themselves at any point necessarily produce modifications in the blood, hence it may be asserted that an examination of the blood furnishes a valuable basis of diagnosis. Yet, from the fact that only blood taken from a superficial vein can be experimented with, and that the blood becomes contaminated in its passage through the body, the small quantity, therefore, of abnormal or noxious matter is often found to be too slight for the determination of its amount, or in some cases even for its detection.

The chemical facts which we possess in regard to the variations of the blood in different diseases are few. It is only known that in such and such states there is a diminution or increase of this or that principle. We are not sufficiently informed as to the genesis of these substances to be able to decide, whether the morbid condition appertans to one organ rather than to another, or whether the disease is due to a given cause or to some other.

The proximate principles of the blood may also seem to increase, without this increase being either real or as great as would appear: this may be due to a diminution in the total mass of blood.

PLETHORA. — Plethora may be due either to an increase in the proportion of globules, or an augmentation in the volume of the blood; therefore we distinguish between globular and sanguinary plethora.

In the former the globules increase.

In sanguinary plethora—that is, in the augmentation of the mass of the blood—the reverse occurs, as the quantity of blood may increase in greater proportion than the globules.

ANÆMIA.—Here also there may be either diminution of the mass of the globules or a diminution in the total amount of the blood.

In the first case, an increase of water and fibrin is noticed in the blood, and often the number of colourless globules increases. The clot is firm and often produces "buff."

The anæmic state occurs when the body does not repair the losses which it has undergone; it is produced during growth, at the time of puberty, or after diseases which impede digestion. Iron and its preparations have a very favourable influence on the development of globules.

We have just stated that certain anæmic conditions correspond to an increase of colourless globules. The spleen then increases in size; the blood which remains in the spleen is very rich in white globules, containing 1 to 49 of the coloured globules. The blood of the splenic vein also contains large numbers of these globules. The coagulum of the blood of this vein is but slightly compact; the serum which separates therefrom coagulates after a short time.

The following hypothesis based upon these facts has been proposed: The spleen is an organ which destroys red globules, changing them into white globules which are carried along into the circulation and afterwards again transformed into red globules. These views are, however, not regarded as established.

LEUCOCYTHÆMIA.—This name is given to a morbid state characterized by the abundance of white globules: the number of these may amount to one-fourth and more of the total number of globules. The blood is then milky, and often acid from the formation of acetic or lactic acid.

CHOLERA—TYPHOID FEVER.—The globules assume irregular forms, and unite together during cholera and typhus. In this latter disease, and in tuberculosis in its advanced stages, the blood loses its property of becoming red in contact with oxygen, since this gas no longer unites with the globules. The blood of typhus patients contains ammonium carbonate, produced by the transformation of urea, and it is probably this compound which leads to the alteration of the globules, as the same phenomenon is observed when ammonia is introduced into the blood.

Ammonia and many toxic agents attack the envelopes of the globules; hence, whenever these substances are present in the blood, the globules become ruptured, and death ensues in the absence of prompt antidotes.

The blood is thick, and resembles gooseberry jelly in

cholera; globules, as well as albumen and extractive substances abound. The serum is deficient, is dense, and generally poor in salts, yet the potassa compounds and phosphates increase. As the urinary secretion is diminished or suppressed, the urea increases in the blood, and there is produced ammonium carbonate.

Scurvy.—The change in the blood is quite marked in this morbid state. It is disorganized on account of the dissolution of the globules, and the diminution of albumen and salts.

Albuminuria.—The blood does not seem to change in the amount of fibrin. The proportion of globules and albumen is greatly diminished.

Dropsy.—The globules and albumen diminish, and the serum is extravasated.

Inflammatory Diseases —The fibrin increases in these affections, in pleurisy, pneumonia, and acute articular rheumatism. The proportion of this body, which, in normal blood, is 2 to 2.3, rises to 7.8 and even 9 parts in 1000.

The fatty matters augment, and the albumen and globules diminish slightly.

The blood is charged with carbon dioxide, which fact explains the retarding of the coagulation, as a large proportion of this gas prevents coagulation.

Diseases in which the Fibrin Diminishes.—When food is insufficient, also in cases of syphilis, in prolonged suppuration, in typhoid fever, and in scurvy, the fibrin generally diminishes, or loses its property of coagulating.

The coagulation of the blood is very slow in diseases of the respiratory organs, when the hematosis is incomplete, and after death by syncope. It does not occur in the blood of persons asphyxiated, killed by lightning, or poisoned with cyanhydric acid, narcotics, hydrogen sulphide, or ammonia. Usually in a fatal result there is a complete destruction of the globules. In this case, oxygen ceased to unite with the blood, and the serum becomes coloured. The blood of persons who have died from the bite of a serpent coagulates very rapidly. It should be remarked that a decrease in the amount of fibrin in the blood does not always occur in the cases as cited above, and, indeed, it is claimed by Gorup-Besanez (21–364) that in no disease whatever is there uniformly a diminution in the fibrin.

VARIATION IN THE ALBUMEN.—The blood becomes poor in albumen under a great many circumstances: after loss of blood, prolonged suppuration, in albuminuria and dropsy, in malarial fevers, in typhoid fever, and scurvy.

The albumen seems to diminish in proportion as the fibrin increases.

VARIATIONS IN ALKALINITY. — Normal blood is alkaline. This alkalinity increases in typhoid and putrid fevers, which is probably due to the formation in the blood of ammonium carbonate from urea.

The blood has been known to become acid after an abnormal production of lactic acid. The globules are then dissolved by this body, and death rapidly ensues.

The alkalinity seems to diminish in inflammatory diseases.

VARIATIONS IN THE FATTY BODIES.—The drinking of large quantities of fluids augments the proportion of the fatty bodies, and it seems certain that corpulent persons would grow thin on diminishing the quantity of liquid which they imbibe.

The fatty matters generally augment during affections of the liver, in phlegmasia, Bright's disease, and in the first stage of some acute diseases.

Z. Pupier (9-80-1146) has lately found by extended researches that the use of sodium bicarbonate or alkaline mineral waters tends to increase the number of red blood-corpuscles both in man and animals.

OTHER VARIATIONS.—The *extractive matters* become abundant in puerperal fever and scurvy.

Claude Bernard recently (9-83-407) set forth the following, based upon his investigations regarding the quantity of *sugar* in the blood.

The sugar of the blood is soon decomposed on the removal of the latter from tne body. After death the sugar also rapidly decomposes, even when retained in the blood-vessels. The presence of sugar is independent of the nature of the food; in the arteries it is uniform in quantity, while in the veins, except in the hepatic, though variable, it is yet less than in the arterial system.

The amount of sugar increases in diabetes.

To extract the sugar of the blood, the latter is first defibrinated. To the serum is added its triple volume of alcohol; the coagulum is separated and washed with water containing an equal volume of alcohol. It is now evaporated to dryness, and the residue treated with alcohol, which dissolves the sugar.

V. Feltz (9-80-553, 1338) recently ascertained by his investigations upon the action of putrefying blood upon animals, that injection of the same into a vein of an animal produced septicæmia. The poisonous properties of putrefied blood are not changed by passing air through the same, but are lessened by the action of pure oxygen. If the gases of the blood are removed with a pump and the blood allowed to remain in a vacuum for some time, it loses its poisonous properties. Feltz is of the opinion that the poisonous body is a gas. In all stages of putrefaction, even after being dried in the air, blood retains the property of producing septicæmia.

Uric acid is sometimes observed to increase in the blood.

The blood of icterical persons contains the colouring matter and other constituents of the bile.

Urea accumulates in the blood when the kidneys perform their functions badly; this condition is known by the name of *uræmia*. The urea which accumulates in the blood is partially decomposed, producing ammonium carbonate.

Von Gorup Besanez (75-23-135) found in the blood of a man suffering with atrophy of the liver, besides the normal constituents, a body closely related to gluten, but very different in its optical properties, hypoxanthin in not inappreciable quantity, formic acid, and volatile fatty acids, rich in carbon, also a non-volatile strong organic acid, soluble in water, alcohol, and ether, which, however, is not lactic acid. Uric acid, xanthin, leucin, and tyrosin could not be found.

The proportion of salts diminishes in intermittent fevers, scurvy, Bright's disease, dysentery, and typhoid states. It augments in intermittent fevers and cholera.

RESPIRATION.

The atmosphere penetrates certain special organs, which are the *lungs* in man, *branchia* in fishes, and *trachea* in insects. There is thus established a continual *exchange* between the blood and the air, which is called *respiration*.

The oxygen of the air coming in contact with the membranous walls of the respiratory organ, which are very thin and very permeable, traverses them and penetrates the blood. It is not dissolved in the serum of this liquid, but it fastens itself upon the globules, and forms with their substance a very unstable combination. Inversely, the carbon dioxide and aqueous vapour on reaching the lungs in the venous blood escape through the same membranes, and are exhaled into the atmosphere to be again shortly decomposed by the green portions of plants.

THEORY OF RESPIRATION.

DIFFERENT methods have been employed for studying the phenomena of respiration. Lavoisier was the first to solve the problem; his method, which has since been perfected by Regnault and Reiset, consists in placing the subject to be experimented upon in a known volume of oxygen, absorbing the carbon dioxide exhaled and renewing the oxygen, in a continuous manner.

A second method consists in placing the subject in a confined space and analyzing this air, determining the volume of gas exhaled at each expiration, counting the number of respirations made during a certain time, and analyzing the air exhaled during this time.

By this method absolute results cannot be obtained, as nitrogen is also exhaled during respiration, and thus we have two unknown data: the weight of the nitrogen exhaled, and that of the oxygen consumed to form water.

Boussingault made use of an indirect method, which consisted in feeding the animal in such a manner that its weight remained constant, also weighing and analyzing the food, as well as the excrements, and subtracting the weight of the latter from the former.

It is clear that the difference between these two weights represents what had been lost by pulmonary and cutaneous respiration.

Boussingault experimented on horses, cows, and doves.

The quantity of oxygen consumed is proportional to the energy with which the vital functions are executed.

Dumas, experimenting on himself, found that the absorption of oxygen was at the maximum 23 litres or 33 grammes per hour, or about 800 grammes for 24 hours; 13 litres of carbon dioxide are produced; the air expired contains 4 per cent. of this gas.

Substantially, the amount of oxygen consumed varies between 20 and 25 litres per hour, or 29 to 36 grammes for an adult man in a state of repose.

We are indebted to Scharling, Andral, and Gavarret, also to Pettenkofer, Regnault, and Reiset for important researches on respiration.

The apparatus of Scharling consists of a chamber of one cubic metre capacity, made absolutely tight by a covering of sized paper. The subject is placed in this for half an hour to one hour. The air enters the chamber through an orifice in the lower portion, and is drawn in by a water aspirator. The products of respiration pass into a series of flasks, the first of which contains sulphuric acid, which retains the moisture, the remainder containing alkaline substances to absorb the carbon dioxide formed.

Two important objections to this method may be stated. The air is not sufficiently renewed, and the chamber is too small. It results, therefore, that the air of the box becomes charged with carbon dioxide and aqueous vapour, and becomes elevated in temperature in an unnatural manner. These circumstances exert a deleterious influence upon respiration, and must necessarily bring about abnormal conditions.

Scharling found that in the respiration of a man 34 grammes or 17 to 18 litres of carbon dioxide are proper hour.

Andral and Gavarret took special care not to effect any modification of the normal conditions of respiration.

A mask of thin copper, the edges of which were furnished with a cushion of caoutchouc in order to prevent any escape of gas, is fixed firmly to the face of the subject, which it covers without binding.

This mask is large enough to receive the product of an entire respiration, and opposite the eyes it is pierced with two orifices closed with glass.

The air penetrates the mask by two tubes, which enter the mask at the height of the corners of the lips. The air expired does not pass out through these tubes, as they contain two little balls of elder-pith, which serve as valves. The air escapes through an opening situated opposite the mouth, and enters into three flasks, from which the air has been exhausted, and whose capacity is 140 litres.

The chief difficulty consisted in regulating the opening of the cock which separates the flasks from the opening in the mask, in such a manner that respiration could take place easily, both for inspiration and expiration.

The operation lasted from eight to thirteen minutes, and the gas collected was about 130 litres.

The cock was closed, the air was permitted to cool in the flasks, and the pressure and temperature determined. Then these flasks were placed in connection with three others exhausted, but separated from the first by tubes arranged for absorbing moisture and carbon dioxide.

The gas was made to pass through the tubes slowly by opening progressively the cocks, and when the gas ceased to pass through the tubes the pressure in the first flask was again measured, the difference giving the amount of air which escaped. The increase in weight of the tubes containing the alkaline solutions represents the amount of carbon dioxide in this air.

The experimenters operated on 37 men and 26 women of various ages, with results which we will now state.

The respiratory phenomena attain their maximum energy at about thirty years of age; they increase up to this age, then decrease until death. From 20 to 30 years the quantity of carbon dioxide exhaled is 18 to 20 litres per hour.

Respiration is more active in men than in women. The production of carbon dioxide is greater during digestion than when fasting; the relation increases from 24 to 33, and even more. At the age of puberty there is a great increase in the production of carbon dioxide in man. This increase is arrested in woman at the age when menstruation sets in, and returns during several years after the critical age. It likewise increases during gestation.

Respiration is feebler during sleep, and, according to Scharling, the quantity of carbon dioxide produced during sleep is one-fourth less than when awake.

Exhaled air contains aqueous vapour; this fact was observed by the ancients, for, on breathing upon glass, or other polished surface, a condensation of droplets of water was observed. This water was considered as exclusively derived from that introduced into the body with the food. Lavoisier distinguished water of *pulmonary transpiration*, proceeding from the lungs, from the *water of respiration* formed by the combination of oxygen with hydrogen.

According to Valentin, the weight of water exhaled from the lungs during 24 hours is, in the mean, 540 grammes, while, according to Barral, it attains to nearly 650 grammes.

It seems certain that expired air removes from the body a small amount more of nitrogen than the air inhaled introduces. According to Edwards, animals absorb nitrogen from the air, and disengage a small quantity of the nitrogen of their own substance. The researches of Regnault and Reiset, however, have demonstrated that the nitrogen of the air is not ordinarily absorbed during respiration, and, consequently, does not assist in nutrition under normal conditions.

Among other principal conclusions of their important investigation were the following :—

1st. When warm-blooded animals are submitted to their habitual alimentary regimen, they always disengage nitrogen; but the quantity of this gas is very small; it never amounts to more than $\frac{1}{80}$ of the weight of oxygen consumed, and is often less than $\frac{1}{150}$.

2nd. When the animals are in a state of inanition they often absorb nitrogen, and the proportion varies between the same limits as that of the nitrogen exhaled in the case where they are subjected to their natural regimen. The absorption of nitrogen almost always occurs in starving birds, but very rarely in mammalia.

3rd. The relation between the quantity of oxygen contained in the carbon dioxide and the total quantity of oxygen consumed seems to depend much more upon the nature of the food than upon the class to which the animal belongs. This proportion is greater when the animals are fed with grain, and in this case exceeds the normal or unity. When they are fed exclusively with meat, this proportion becomes less, and varies from 0.62 to 0.80.

With a diet of vegetables, the relation is in general intermediate between the two just given.

4th. The relation between the oxygen contained in the carbon dioxide and the total oxygen consumed varies for the same animal from 0.62 to 1.04, according to the diet to which it is subjected. It is therefore far from being constant.

5th. The quantities of oxygen consumed by the same animal in equal times vary much, according to the different periods of digestion, the amount of activity, and many other circumstances. With animals of the same species, and of the same weight, the consumption of oxygen is greater in young than in adults; it is greater in lean healthy animals than in very fat ones.

6th. Warm-blooded animals disengage by respiration small and almost indeterminable quantities of ammonia and sulphuretted gases.

7th. The respiration of animals of different classes in an atmosphere containing two or three times as much oxygen as normal air presents no difference from that which takes place in our terrestrial atmosphere. The consumption of oxygen is the same; the relation between the oxygen contained in the carbon dioxide and the total oxygen consumed undergoes no perceptible change; the proportion of nitrogen gas exhaled is the same, and the animals do not seem to perceive that they are in an atmosphere different from the ordinary one.

In the recent experiments of Bert, he observed that if an animal be exposed to the influence of pure oxygen under a pressure of four atmospheres, it gives signs of discomfort, which are followed by violent convulsions, and death ensues if the pressure be increased to five atmospheres.

It is to the action of the oxygen and not to the increased pressure that these effects are to be attributed; for if a swallow be exposed to air under a pressure of three atmospheres, and then nitrogen at twenty atmospheres admitted, the animal perishes, slowly asphyxiated, without convulsions. The convulsions also ensue if the oxygen under four atmospheres pressure be replaced with air under twenty atmospheres. The analysis of the gases of the blood shows that when death ensues the blood, instead of containing 18 to 20 c.c. of oxygen in 100, as in ordinary conditions, contains 35 c.c. An unusual combustion does not take place, for the temperature of the animal seems to fall sensibly, or at least it does not increase.

On the other hand, when the pressure of the air is diminished until death ensues, the bird perishes asphyxiated in the midst of a pure air hardly containing any carbon dioxide. Then death takes place because the pressure of the oxygen is not sufficient to maintain in the blood the quantity necessary for producing vital phenomena.

Thus an aeronaut would be able to mount without danger to much greater heights than have hitherto been reached if he would inhale oxygen when suffering from the rarefaction of air.

On the other hand, divers would be able to work at great depths without danger if, instead of sending them pure air, a mixture of air and nitrogen of definite proportions were supplied.

SUMMARY OF THE THEORY OF RESPIRATION.

Priestley was cognizant of the fact that air and oxygen—which latter element he had just discovered—had the property of reddening venous blood, and that carbon dioxide turned arterial blood to a brown colour. But, misled by the phlogistic theory, he did not have the satisfaction of establishing the theory of respiration and combustion—an honour which belongs entirely to Lavoisier.

This theory enabled the latter chemist to explain animal heat, and in 1789 he wrote the following:—

" Respiration is simply a slow combustion of carbon

and hydrogen which is similar, on the whole, to that which takes place in the flame of a candle. Animal organisms are thus true combustibles, as they are oxidized in respiration and consumed at the expense of the oxygen of the air."

As to the part of the body in which the combustion took place, he did not claim to make any assertion.

Lagrange was the first to state that combustion takes place in the capillaries, and since then many investigators have established, by experiment, the truth of this assertion.

In order to show, however, that the change in the colour of the blood takes place in the lungs, we have only to observe the lungs of a frog after having, by appropriate dissection, exposed them to view. The transparency of the membranes admits of the difference in the colour of the blood being plainly seen before and after leaving the lungs.

The air produces this change, for if, as was done by Bichat, a cock be adapted to the carotid artery of a dog, the blood, which is red, becomes black when air is prevented from entering the lungs by closing a cock placed in the trachea; and the red colour returns as soon as air is allowed to enter.

The fact, perfectly demonstrated above, in regard to the relative insufficiency of oxygen and the abundance of carbon dioxide in the venous blood as compared with arterial blood, proves indirectly that an absorption of oxygen and a production of carbon dioxide takes place in the capillaries situated between the arteries and veins. But Spallanzani, and especially W. Edwards, have directly proved this important fact. The latter removed the gases from the lungs of a frog by compressing them under mercury, and introduced the animal under a bell-glass filled with hydrogen over mercury. The frog breathed for quite a long time; the analysis of the gases showed that they contained a volume of carbon dioxide much greater than that exhaled by the animal under ordinary conditions.

OXYGEN.—Oxygen is not merely dissolved by the blood. If such were the case, the blood of persons living in mountains would contain less than that of those who live in the lowlands. Nothing of this kind has been observed at Quito (2,908 metres above the level of the sea), at Potosi (4,166 metres), or at Deba (4,812 metres); in the latter place, the atmospheric pressure is scarcely half of that at the level of the sea, and consequently the blood should contain but half the amount of oxygen. On the other hand, Regnault and Reiset have observed that the absorption of oxygen does not increase when animals respire an atmosphere containing two or three times as much oxygen as ordinary air.

It is well known that the quantity of any gas dissolved is directly proportional to the pressure which the gas sustains.

On the other hand, the coefficient of solubility of oxygen in the blood at 15° is 0.0287, or very nearly that of oxygen in water; and it is the same for serum. Hence it results that one litre of blood should dissolve only $\frac{0.0287}{5}$ of oxygen, or 5.7 c.c., while the real amount contained in the blood is 92 to 95 c.c. (Fernet).

It is therefore probable, à *priori*, that oxygen forms a combination with one of the principles of the blood. This principle is not the serum; for if the blood be defibrinated and the globules removed, the serum dissolves scarcely more oxygen than water. If, on the contrary, defibrinated blood containing the globules be agitated with oxygen, it absorbs much more oxygen than the serum deprived of globules. If the globules simply dissolved the oxygen, the proportion would increase as the temperature decreased; but this is not the case. At 40° to 45° a maximum absorption is observed, and at a higher temperature the phenomena of oxidation take place. A combination is therefore produced, and it follows from what we have said above that the hæmoglobin of the globules must be the agent which effects the combination with the oxygen. This combination is also extremely unstable, as the oxygen may be almost completely removed in a vacuum.

The oxygen of the blood acquires an energetic oxidizing power, comparable to that of ozone, at a temperature where ordinary oxygen is inactive. In fact, essence of turpentine, to which a few globules of arterial blood are added, turns litmus at once blue, in the same manner as when agitated in the air in the sunlight. Hydrogen peroxide dissolves pyrogallic acid without becoming coloured; but if to this solution platinum black or blood globules be added the brown coloration is at once produced.

THE BLOOD AND CARBON DIOXIDE. 313

The blood contains, besides oxygen combined with the globules, a small proportion of this gas dissolved in the serum.

It should be also stated in this connection that a small portion of the oxygen inhaled is employed to oxidize the sulphur of complex sulphur compounds, the albuminoids, etc.

CARBON DIOXIDE.—Carbon dioxide is not, like oxygen, combined with the globules. Fernet has determined the quantity of this gas which the constituents of the serum—water, carbonate, phosphate, and chloride of sodium—are capable of absorbing, either by dissolving or combining with it; and the result of these researches shows that the quantity of carbon dioxide found in the blood is very nearly equal to that which the serum alone absorbs.

The quantity of carbon dioxide exhaled is less during sleep than when awake, for, as the organs are at rest, the oxidation is not then so great.

The carbon dioxide is not derived from the atmosphere, since the gases exhaled contain more than the air, and its proportion, which is small in arterial blood, is observed to increase as this liquid traverses the organs in which the combustion takes place. This gas is therefore formed in the body, and is rejected as a waste product.

F. M. Raoult (9–82–1101) finds, as a result of recent experiments, that the presence of carbon dioxide in inhaled air causes a diminution of the carbon dioxide exhaled, and therefore in the oxygen consumed.

NITROGEN.—Nitrogen forms at the most one-tenth of the gases of the blood, which contains 2 to 3 per cent. of this gas, while the serum dissolves only 1 per cent.; consequently there is for this gas a special action not as yet explained.

To review, the vesicles of the lungs act as a porous membrane; and this organ should be regarded as an apparatus for the exchange of gaseous bodies.

The blood which has become red in the lungs retains this colour until it enters the capillaries. On leaving the capillaries it is darker, and instead of oxygen it contains carbon dioxide. Consequently it is in this transit that the combustion takes place. This combustion either occurs in the capillaries proper, or the oxygen traverses their dialyzing walls and penetrates into the depths of the tissues whence the carbon dioxide escapes. This latter hypothesis is more in favour than the first. There is an exchange of gases in the centre of these structures as in the lungs, and the oxygen coming from the air penetrates into the innermost parts of the body of animals, and there effects the oxidation of the tissues themselves.

VARIATIONS IN THE GASES EXPIRED IN PATHOLOGICAL STATES.

We have but little information on this point. According to Hervier and Saint-Lager " the proportion of carbon dioxide decreases in all diseases in which

GASES EXHALED IN PATHOLOGICAL STATES.

respiration is impeded, as in pulmonary phthisis, pneumonia, pleurisy, pericarditis, eruptive fevers, and typhoid affections."

In diabetes, chlorosis, anæmia, and in diseases in which there is no febrile movement, the variations in the proportion of carbon dioxide are hardly appreciable. In inflammations the carbon dioxide increases in a remarkable manner.

Rayer, and afterwards Doyère, have affirmed that the air exhaled by cholera patients contains more oxygen and less carbon dioxide than the air normally expired. The quantity of oxygen absorbed is always greater than that of the carbon dioxide exhaled.

NUTRITION.

ANIMALS cannot live unless able to respire and obtain nourishment, *i.e.*, to ingest matters which are digested, absorbed, transported to the blood and submitted, subsequently, to the action of oxygen.

The food, carried by the blood into the different organs, undergoes therein two different changes. One part is burned, as coal in the furnace, producing heat and physical energy. The remainder becomes organized to form the tissues, since an animal, considered even in an adult state, and at a period at which its weight does not vary, constantly fixes matter in its organism, and therefore also loses an equivalent amount.

ANIMAL HEAT—MUSCULAR POWER.

These two subjects are intimately connected with one another, and with respiration.

The temperature of animals, and even that of plants, is not uniformly that of the medium in which these beings live. It varies also with the species. In man it is very nearly 37 degrees, in whatever climate he may live.

The two extremes of temperature in which man can exist are very remote. He alone is capable of dwelling in all latitudes, in the most varied climates, and at heights so great that the pressure is only one-half of that at the level of the sea.

Different portions of the body have not the same temperature. The exterior parts, from the cooling effect of the surrounding medium, are reduced in temperature 4 to 5 degrees below that of the interior. The muscles are 1.5 to 2 degrees warmer than the cellular tissue.

It is the blood which, traversing the whole body, tends to equalize the heat disengaged in the different organs. The liberation of gases in the lungs lowers their temperature slightly, and especially that of the left cavities as compared with the right. The venous blood in the extremities is slightly less warm than the arterial blood, but this is due to the external position of the vessels.

The conditions which cause the activity of the respiration to vary, that is, the absorption of oxygen, produce also a corresponding variation in animal heat. The temperature of the body of an infant or an old man is less than that of an adult, and we have observed that the respiratory phenomena diminish in energy at the two extreme points of life.

If an important reduction in temperature is produced after eating, it must be attributed to the fact that the blood rushes to the muscles of the digestive apparatus, which act with increased energy at this time.

Like the fuel of an ordinary engine, a part heats the animal machine, the other is converted into muscular activity, which produces either external work (walking, movements of the arms, head, etc.), or internal work (digestion, assimilation, etc.). Thus the observed heat is equal to the difference between the heat produced and the heat which is transformed into work. Now, since we know the mechanical equivalent of heat, that is, the quantity of work which a certain amount of heat will accomplish, the heat produced can be measured.

If the muscle contracts without producing mechanical effect, the heat developed will be greater, since there is only heat developed, and that not utilized in the form of work. But even if the muscular power does produce an external mechanical effect, there is still in addition a production of heat in the interior of the body. Experiment has shown that when a muscle contracts the quantity of oxygen consumed is greater than when it is in repose: thus 100 volumes of blood, leaving a muscle which is in action, instead of furnishing 6 volumes of oxygen, furnish only 2 volumes.

All chemico-physiologists are in accord in admitting that heat and motion are due to the oxidation of the food. The amount of carbon dioxide exhaled does not indicate the amount of oxidation which has taken place in the body. Every movement, every chemical action, every passage of the food from a solid to a liquid state in the blood, all friction of the liquids in the body, are actions which go to produce an elevation or decrease of temperature. Consequently there are incessant gains and losses of heat, and we perceive, on the whole, only the resultant of these different actions of which the complexity is extreme.

The carbon dioxide is not the only product of oxidation; water and other matters (urea, uric acid, etc.) are formed, which escape in the different excretions. And the whole of the oxygen which oxidizes is not derived from the air; a considerable part is obtained from the oxygen of the food itself.

There is a difference of opinion as to the manner in which the action is produced. According to some, it results from the oxidation of the aliments as they are found in the blood. Others do not admit that the process takes place in the blood, but that it is a direct oxidation of the muscles by the oxygen which produces heat and motion.

The second view is that most generally admitted; nevertheless, the recent researches of Meyer and Frankland on this subject appear to prove the contrary.

An average man has about 7.5 kilos of muscles, considered in a dry state. According to Meyer, they would be completely oxidized in eighty days if they served to produce mechanical work.

It is rational to regard the muscles as instruments for the transformation of potential energy into motion. We can only give a few conclusions deduced from the work of Frankland.

1st. The muscle is a machine destined to convert potential energy into mechanical force.

2nd. The mechanical force of the muscles is derived principally, if not wholly, from the oxidation of the substances contained in the blood, and not from the oxidation of the muscles themselves.

3rd. In man the principal substances employed in the production of muscular power are non-nitrogenous; but nitrogenous substances may also be employed for the same object, hence the great increase in the evolution of nitrogen under a diet of animal food, even with no increase in the amount of muscular work performed.

4th. Like all other parts of the body, the muscles are constantly being renewed; but this renewal is not apparently more rapid during great muscular activity than during comparative repose.

5th. After a sufficient quantity of albuminous substances has been digested for the renewal of the tissues, the best food for the production of work, both internal and external, are the non-nitrogenous substances, such as oil, fat, sugar, starch, gum, etc.

6th. The non-nitrogenous portions of the food which enter into the blood transform all their potential energy into effective force; the nitrogenous substances, on the contrary, leave the body, taking with them a part (one-seventh) of their potential force.

7th. The transformation of dynamical force into muscular power is necessarily accompanied by a production of heat within the body, even when the muscular force is exerted exteriorly. This is, without doubt, the principal though not the only source of animal heat.

Fick and Wislicenus, in an ascension of the Faulhorn in 1865, determined the amount of work performed by their muscles, and the quantity of muscular matter oxidized to produce this work. This latter calculation was made by determining the amount of nitrogenous matters in the urine emitted, and collecting them in the form of urea, and based upon the fact established by Frankland, that 1 gr. of dried muscle transformed into urea produces 4,368 heat units. They arrived at the result that the work accomplished was about twice as great as that which would be produced by the combustion of the substance of the muscles transformed into urea.

TRANSFORMATION OF FOOD IN THE BODY.

We recognize three principal classes of food, albuminoid, farinaceous, and fatty.

Transformation of Albuminoid Substances.— It was formerly believed that albuminoid matters were not modified in the body, but simply fixed in the tissues, and taking no part in the respiratory phenomena. The name of *plastic* food given to these bodies illustrates perfectly this manner of regarding them. On the other hand, the fatty and farinaceous bodies were thought to take part in the production of the respiratory phenomena alone. Hence the name *respiratory* aliments, which has been given them. This view, however, is too limited; carbon dioxide and water are the principal but not the only products exhaled. Others are formed, as urea, uric acid, and these substances are nitrogenous. There escapes also in the gas exhaled by the lungs a certain quantity of free nitrogen. It is well known that the framework of animal tissues is nitrogenous; but it is none the less certain that different tissues are filled with non-nitrogenous matters, such as the fat of adipose tissue and glycogenous substances.

The albuminoid matters undergo in the blood, and afterwards in the organs to which the blood carries them, numerous transformations, most frequently produced by oxidation. To prove this it will only be sufficient to enumerate the different nitrogenous principles found in the body. These are mainly:—

Urine	Urea . . .	CH_4N_2O
	Uric acid . .	$C_5H_4N_4O_3$
	Hippuric acid .	$C_9H_9N_4O_3$
	Cystin . .	$C_3H_7NSO_2$
	Xanthin . .	$C_5H_4N_4O_2$
Perspiration—	Sudoric acid .	$C_{10}H_8O_{15}N$?
Liver . . .	Taurocholic acid	$C_{26}H_{45}NO_7S$
	Glycocholic acid	$C_{26}H_{43}NO_6$
	Cholesterin .	$C_{26}H_{44}O + H_2O$
Pancreas .	Leucin . .	$C_6H_{13}NO_2$
	Tyrosin . .	$C_9H_{11}NO_3$
	Lactic acid .	$C_3H_6O_3$
Muscles .	Creatin . .	$C_4H_9N_3O_2$
	Creatinin .	$C_4H_7N_3O$
	Inosite . .	$C_6H_{12}O_6 + 2H_2O$
	Inosic acid .	$C_5H_8N_2O_6$?
	Sarcosin . .	$C_3H_7NO_2$
	Sarcin (Hypozanthin)	$C_5H_4N_4O$

Osseous tissue—Ossein.

TRANSFORMATION OF AMYLACEOUS OR FARINACEOUS FOOD.—Starchy matters are only found in small quantities in the tissues of the body—a fact which is quite natural as regards carnivorous animals, but very surprising in the case of herbivorous animals; and which seems to prove that starch is the chief respiratory aliment, and that it is very easily oxidized or burned. The greater part of the starch is transformed into carbon dioxide and water. Another portion is converted into fat, and the rest (a minute fraction) is fixed in certain tissues.

GLUCOSE IN THE LIVER.—The existence of amylaceous matter in animal tissue is connected with a remarkable discovery made in 1849 by the illustrious physiologist, Claude Bernard—a discovery which we shall now describe, as well as the researches which led to it.

If a carnivorous animal be subjected to prolonged fasting, sugar will be found in the hepatic tissue. The proportion of sugar found in the liver of carnivorous animals, or of animals fed exclusively with meat, is substantially the same as that in the liver of herbivorous animals, or of animals fed with amylaceous or saccharine food. Hence the production of sugar does not depend upon the existence of amylaceous and saccharine substances in the food.

Objections might be raised to such experiments, on the grounds that the blood, in passing through the liver, might leave sugar behind it in this organ, and that sugar is merely retained and accumulated by the liver.

Bernard responds to these suppositions by an experiment as interesting as it is conclusive. If a dog is killed and the liver removed, and, after washing this organ in such a manner as that all the sugar shall be dissolved, it is allowed to remain exposed to the air for a day, it is found to again contain a very large proportion of sugar.

If, also, the blood of the vena porta be analyzed before it reaches the liver, as well as after leaving this organ in the superior hepatic veins, a considerable increase in the amount of sugar is observed. In order to extract the sugar of the liver, the latter is cut into very small pieces and treated with boiling or even with cold water till nothing more is dissolved. The liquid is decoloured with animal charcoal, and evaporated over a water bath almost to dryness, and the residue treated with alcohol. The alcoholic solution furnishes glucose on evaporation.

Bernard found 23.27 gr. of sugar in the liver, weighing 1,300 grammes, of a hanged criminal of forty-three years, and 25.70 grammes in that of another, aged twenty-two, and whose liver weighed 1,200 grammes.

Glycogene.—Sugar is produced in the hepatic tissues by means of a third substance—a sort of animal starch, designated *glycogene*—which has also been found on the internal surface of the amniotic membrane of ruminants, between the maternal and fœtal placenta of rodents, in the muscles, and in the lungs of the fœtus, and later in the liver; also in different parts of the crustacea and articulates.

GLUCOSE IN THE LIVER.

To prepare glycogene, the liver of a dog recently killed is cut into very small pieces and thrown into boiling water to precipitate and destroy the ferment which would otherwise change the starch into sugar. The fragments are now withdrawn, triturated with animal charcoal, and the pulp obtained boiled for about twelve minutes with five times its weight of water, filtered, and the residue treated with additional water. A liquid is obtained, from which the glycogene may be precipitated by alcohol.

Glycogene is a white powder, soluble in water, which it renders milky, and insoluble in alcohol. The solution turns the plane of polarization strongly to the right. It has the composition of starch, $x\,(C_6H_{10}O_5)$, is coloured violet-red by iodine, is converted into pyroxam by fuming nitric acid, and furnishes dextrine and glucose under the same circumstances as vegetable starch.

The transformation of glycogene into sugar is effected by means of a ferment analogous to diastase, which is found in fresh liver and even in the blood.

According to Pavy, the proportion of glycogene in the liver varies with the nutrition; it is large if the food is vegetable, and is, on the contrary, small if the food is animal.

	Amount of Glycogene in the Liver.
Dog fed with amylaceous food.	17.23 per cent.
,, ,, ,, meat.	6.97 ,,
,, ,, ,, ,, mixed with sugar	14.50 ,,

Rouget arrived at analogous results. On the other hand, Sanson has announced that on giving animals very farinaceous food, dextrin is found in the blood and even in the muscles; consequently muscles supplied as food would furnish amylaceous matters directly. There also exists in the muscles a saccharine substance called *inosite*, $C_6H_{12}O_6$, and lactic acid; consequently a diet of meat forms in the body amylaceous products.

Amylaceous matter is also found in the muscles of new-born mammalia, and in the muscles of an organ when in absolute repose for a certain time. This all leads to the belief that there is an amylaceous matter which takes part in the formation of muscular tissues, but which disappears under ordinary circumstances, and is transformed into inosite and lactic acid.

From these facts, and the existence of glycogene in other parts of the body than the liver, it follows that the liver is not absolutely the only organ having the property of transforming starch into sugar, but that it possesses it in a much greater degree than do the others.

The sugar thus formed in the liver then passes into the blood, and there disappears, under normal conditions, being burned by the oxygen; but certain natural or artificial conditions may diminish or increase the formation of this sugar.

If the spinal cord be dissevered below the phrenic nerves, the circulation becomes weaker in the abdominal region, the temperature is lowered, and sugar is no longer found in the hepatic veins.

ARTIFICIAL AND NATURAL DIABETES.

It is observed that the amount of sugar increases in the blood of the superior hepatic veins when the pneumo-gastric nerves are irritated, when a special point in the wall of the fourth ventricle is pricked, when essence of turpentine, ether, or chloroform is injected into the vena porta, or simply when large proportions of these agents are inhaled, or, finally, when poisoning is produced by curarina, strychnia, or brucia.

Let us follow step by step the research of Bouchardat, in order to study the theory of natural diabetes. And first we will recall the fact, that the digestion of amylaceous substances takes place in the intestines under the action of the pancreatic and intestinal juices, that the greater part of the starch is only changed into dextrine in the intestine, and that the further transformation of this dextrine takes place chiefly in the blood, under the action of the intestinal diastase absorbed simultaneously with the dextrine.

Whenever there is an excess of glucose in the blood, this sugar passes into the urine. This fact may be demonstrated by injecting glucose into the veins: if there is but little, none is found in the urine; if there is a large amount present, reagents will indicate its presence in the urinary secretion.

The causes which produce an excess of glucose in the blood may be of two opposite characters: either the sugar is due to too great a secretion, or it may result from an insufficient destruction; but more often veritable glycosuria characterized by a constant excess of sugar, is due to both of these causes combined. It has been demonstrated that the sugar passes into the urine whenever there is more than 3 to 5 grammes in the blood at one time.

There may be an incompleteness in the destruction of the glucose in the blood, either because the oxygen is not present in sufficient quantity or because it meets with substances which are more easily oxidized.

Diabetes will result when, the nutrition being very starchy, there is an excessive transformation of amylaceous substance into glucose in the digestive canal. In fact the glucose is observed to increase with the proportion of amylaceous food. In persons affected with glycosuria the transformation takes place in the stomach, and this fact consequently explains why all albuminoid substances are susceptible of acting upon starch; they differ only in the rapidity of their action. It has also been shown that if the pancreas of a pigeon be removed it will still be able to digest amylaceous substances. Bouchardat has also observed that the stomach of persons having glycosuria is generally very much enlarged, and that persons who have a tendency to diabetes prefer farinaceous food, that they eat a great deal, and also that they eat rapidly, which circumstances occasion a longer sojourn of the food in the stomach. When an organ is much used it acquires greater strength, and it is not unreasonable to admit that under these circumstances the gastric juice may not be sensibly changed, and become incapable finally of dissolving amylaceous matter.

Diabetes is accompanied by continual thirst; hence it will be understood that since the food requires 8 to 10 times its weight of water for digestion, the gastric juice must be insufficient if the digestion of the farinaceous food takes place in the stomach at the same time as the albuminoid.

The sugar in the urine of diabetic persons ordinarily disappears on submitting them to a diet formed exclusively of meat, if the disease is not too advanced.

In general, any cause on the other hand which produces a diminution of the respiratory phenomena tends to retard the destruction of glucose in the blood and produce diabetes if the tissues are saturated with glycogenic matters.

TRANSFORMATION OF FATTY SUBSTANCES.

It has been established by a large number of experimenters, who have operated upon different animals, that all of them not only assimilate fatty matters, but that they produce fat as well. Fat alone given as food produces inanition. If animals be submitted to varied nutrition, there is much more assimilated fat found than there was in the food originally supplied them. Fatty bodies mixed with the other food facilitate growth. Amylaceous and saccharine substances are readily changed by digestion into fatty matters. It has not been demonstrated that nitrogenous foods are transformed into fats.

The Rôle of Mineral Compounds in Nutrition is but little understood. *Iron* exists in different parts of the body, and principally in the blood globules.

Sodium chloride is found in most animal fluids. It is thought, as we have already stated, that this substance is the origin of the hydrochloric acid of the gastric juice, and of the soda, which is found in the intestinal juices. It is known that this salt forms a compound with glucose (p. 186), also the existence of a compound of sodium chloride and urea has been shown; and this is the reason for the belief that salt assists in the transformation and elimination of sugar and of urea. It aids in the solution of albumen and casein in certain humours. It prevents the dissolution of the blood globules, of the chyle and lymph, and we have reason to believe that it, like other salts elsewhere, is an important factor in the absorption of liquids by different membranes.

Weiske and Wildt (7-1874-123) have made investigations as to the action of food poor in lime and phosphoric acid, upon animals of rapid growth. They experimented upon three lambs about two and a-half months old, and in a healthy condition, feeding one with food poor in lime compounds, one with food poor in phosphoric acid, and the third with the usual kind of food; while the latter prospered and gained 13.5 pounds in fifty-five days, the first two lost thirteen and fourteen pounds in weight, and were by this time nearly dead. The animals having been killed, the composition of their bones, as regards their inorganic constituents, were alike, but the amount of fat in the bones of the animal fed with normal food was greater than in both the others. A diet poor in calcium and phosphorous compounds does not affect the constitution of the bones as regards their mineral constituents.

Sodium Phosphate is capable of facilitating the absorption of carbon dioxide by the blood, and consequently it is regarded as playing an important part in respiration.

Calcium Phosphate is found in the majority of animal substances. This salt forms the greater part of the mineral matter of the bones, it exists in the ash of albuminoid compounds. It enters the body dissolved in water by means of carbonic acid.

This substance, as well as the calcium carbonate, magnesium phosphate, and silica assist in giving solidity to the animal structure, and Chossat has asserted that the bones of pigeons completely deprived of calcium phosphate become so thin as to break. Magnesium phosphate cannot replace calcium phosphate.

Weiske (36-'77) has investigated the influence of common salt upon the live-weight and the disassociation of nitrogen in various animals, and ascertained: that if the amount of salt in the food increases, and the animal be allowed all the water it desires, the amount of water consumed increases; that with the increase of salt in the food and the consumption of water, as far as an increase in the production of urine accompanies the same, the disassociation of nitrogen increases: that when the salt is removed, the consumption of water, as well as the production of urine, and disassociation of nitrogen, decreases; nevertheless the latter remains higher for a longer time than if a large ingestion of salt had not taken place. The increase in weight following a diet composed largely of salt is not due to increase in the amount of flesh, but to the accumulation of water in the body. Salt given in the food increases the desire for eating, but a notable increase or decrease in the digestibility of the food has not been proven.

URINE.

HUMAN urine in its normal state is a liquid of an amber colour, the concentration of which, and consequently the density, varies with the age, sex, and state of digestion. This secretion is much more abundant, relatively, in infants than in grown persons, but the urine of infants is also richer in water, paler and less dense than that of adults. Parrot and A. Robin have lately (9-82-104) studied the urine of newly-born infants, and find that the secretion amounts to four times as much, referred to the weight of the body, as in adults.

The quantity of urine in woman is to that in man nearly in the proportion of 13 to 12.

The urine of man is pale, and charged with water after abundant ingestions of this liquid. Normal urine is that obtained soon after rising, its density is about 1.018, it varies between 1.012 and 1.022; its density may fall as low as 1.003, and rise to 1.030 after a hearty repast; it is then yellow.

Water is evacuated from five to six hours after having been taken into the system. The proportion of urine is extremely variable; 1,200 to 1,300 grammes is about the mean in men in twenty-four hours; 1,300 to 1,400 grammes in women. But this quantity may sometimes increase to 2,000 grammes, and descend to 900 grammes.

The three principal causes which influence the amount of this secretion are:—

1st. The nature of the blood; a very aqueous blood increases it.

2nd. The rapidity of circulation in the kidneys.

3rd. The activity of the pulmonary and cutaneous respiration. The urinary secretion varies in inverse proportion to the respiratory phenomena; thus the quantity of urine emitted is greater in winter than in summer, in cold countries than in warm countries. After a cold bath the urinary secretion attains its maximum.

Certain salts—nitre, for example—increases the quantity of urine; they are denominated diuretics.

Other substances retard and diminish this secretion, as cantharides, etc.

The proportion of solids extracted from the body by the urine may vary from 40 to 80 grammes in twenty-four hours.

Composition of Normal Urine of Man.

Water.	936.76	931.42	932.41
Solid constituents.	63.24	68.58	67.59
	1000.00	1000.00	1000.00

The solids are composed of—

Urea	31.45	32.91	32.90
Uric acid	1.02	1.07	1.07
Lactic acid	1.49	1.55	1.51
Aqueous extract	1.62	0.59	0.63
Alcoholic extract	10.06	9.81	10.87
Lactate of ammonium	1.89	1.96	1.73
Chloride of sodium and of ammonium	3.64	3.60	3.71
Alkaline sulphates	7.31	7.29	7.32
Sodium phosphate	3.76	3.66	3.98
Calcium and magnesium phosphates	1.13	1.18	1.10
Mucus	0.11	0.10	0.11
	63.48	63.72	64.90

(Lehman.)

INFLUENCE OF THE FOOD ON THE COMPOSITION OF THE URINE.

Nature of the Food.	Solids in 1000 parts.	Urea.	Uric Acid.	Lactic Acid and Lactates.	Extractive Matters.
Honey	67.82	32.498	1.183	2.725	10.489
Animal	87.44	53.198	1.478	2.167	15.196
Vegetable	59.24	22.481	1.021	2.669	6.499
Non-nitrogenous	41.68	14.408	0.735	5.276	11.854

(Lehman.)

Normal human urine is acid. This acidity is due to the action of the uric acid and other acids of the urine upon the alkaline phosphates. These acids deprive them of a portion of their alkali, and acid phosphates result. Uric or hippuric acid may also be found in excess in urine.

The quantity of free acid evacuated in twenty-four hours represents 2. to 2.5 grammes of oxalic acid.

The reaction of the urine depends upon the character of the food. In fact, this secretion is alkaline in herbivorous animals, since their food, which is very rich in carbon, forms bicarbonates with the bases which are in this secretion; but the urine of an herbivorous animal may be rendered acid on submitting it to a diet of flesh food. The urine of herbivorous animals is turbid, and contains urea, hippuric acid, and a small quantity of phosphates; it does not contain uric acid.

Inversely the urine of carnivorous animals is acid and clear. It is rendered alkaline by forcing the animals to an exclusive vegetable diet.

The urine of carnivora contains more urea and uric acid than that of man or herbivorous animals, while hippuric acid is wanting in it. Regarding the occurrence of phenol, E. Bauman has recently observed that albumen and pancreas in putrefying form a certain quantity of phenol, and he believes in this reaction can be found an explanation of the existence of phenylsulphates in the urine of dogs fed exclusively with meat (60-'77-685).

Violent exercise, fatigue, and excesses render human

urine alkaline. This fact is due to the combustion which, under these circumstances, transforms the uric acid into urea, and this body does not possess, like uric acid, the property of removing from the phosphates a portion of the alkali which they contain.

Gosselin and A. Robin (9-78-72) have made experiments upon animals, injecting ammoniacal urine subcutaneously, and found that animals subjected to this treatment became feverish, and when larger quantities were injected they died. Thus in diseases of the bladder, the ammoniacal urine, if reabsorbed, must be deleterious, hence it would be advantageous to the patient that the amount of ammonium carbonate in the urine be reduced; this, according to investigations of Gosselin and Robin, is effected by the administration of benzoic acid. Pasteur (9-78-46) claims that the ammoniacal nature of urine is due to the action of a ferment which obtains entrance through the urinary passages, or sometimes is introduced mechanically by means of chirurgical instruments. He recommends, therefore, that the instruments before being used be plunged into boiling water, or heated, then quickly cooled, and at once employed.

A. Lailler (9-78-361) is of the opinion that the ammoniacal fermentation of urine depends in a great measure on the amount of mucus it contains.

Gubler (9-78-1054) asserts that the decomposition of urea into ammonium carbonate, as is the case in the bladder in certain diseases of this organ, is due to small pus-corpuscles (*néocytes*).

W. Zuelzer (60–1875–1670) has lately found that after a diet composed wholly of meat, the urine of a dog contained for every 100 parts by weight of nitrogen, 12 to 14 of phosphoric acid; when fed with potatoes and bread, it contained 20 to 30 of phosphoric acid to 100 of nitrogen. In a healthy man, 20 to 25 years of age, the food being mixed and sufficient, the urine contains 17 to 19 of phosphoric acid to 100 of nitrogen; with a diet of meat the proportion of phosphoric acid decreases, with a vegetable diet it increases. The time of day and the state of health have great influence upon the relative proportion of these two substances. Under normal conditions a man eliminates 12 to 14 of sulphuric acid, 0.3 to 0.7 of lime, and 0.6 to 1.0 of magnesia to 100 of nitrogen.

The urine, on leaving the body, deposits mucus after a certain time: it often also deposits urates, especially during fevers. But its acidity soon increases, in consequence of the formation of more uric acid; this acid is often seen to deposit in the form of rhomboidal prisms. Other acids are also formed, chiefly acetic and lactic acids. At the end of a few days the urine loses its acidity and becomes decidedly alkaline from the formation of a considerable quantity of ammonium carbonate. This salt is formed from the urea thus:—

$$\left. \begin{array}{c} CO'' \\ H_2 \\ H_2 \end{array} \right\} N + 2H_2O = 2(NH_4) \left\{ \begin{array}{c} CO'' \\ O^2 \end{array} \right.$$

This transformation of urea is favoured by the

presence of the mucous sediment which urine deposits when exposed to the air, also by the action of beer, yeast, and albuminoid substances.

It is a true fermentation, accompanied by the development of an organized vegetable substance (*Torulaceæ*), which reproduces itself by germination. Often its action is impeded by the formation of infusoria, which maintain the acidity of the urine for a long period. Cohn finds the organisms to be *Micrococus ureae*.

NORMAL CONSTITUENTS OF THE URINE.

Of the solid constituents, urea is the most abundant. The urinary secretion in man furnishes about 30 grammes of urea in 24 hours, but this quantity may vary greatly. The average in women is 20 grammes; it falls to 9 grammes in old men. A very nitrogenous diet increases it, while food which is poor in nitrogen diminishes it. Urea does not even disappear in an animal rigorously kept without food; it is then formed at the expense of the tissues.

When the urinary secretion increases, even though from the drinking of large quantities of water, the amount of urea produced also increases. It augments likewise, according to some authorities, during severe physical labour.

We may admit, in general, that urea diminishes when the circulation of blood is sluggish, and that it increases when the circulation becomes active.

There is only a very small quantity of urea in the

blood; it becomes greater when the kidneys perform their functions badly. Urea is not formed in the kidneys. Dumas and Prevost showed in 1823 that the blood of animals, from which these organs have been removed, contains considerable amounts of urea.

This fact has been confirmed by Bernard and Barreswil, who also showed that, after the removal of the kidneys, the gastric and intestinal secretions increase. The gastric juice remains acid but contains ammonia. When the animal becomes entirely exhausted, urea is found in the blood in a very notable quantity.

Picart and Meissner have obtained the same results, which have, however, been doubted by Oppler, Perls, and Zalesky. The question has been taken up by Grehant, who conceived the idea of determining the amount of urea with the greatest care, and he has perfectly demonstrated that urea accumulates in the blood in consequence of nephrotomy.

100 grammes of arterial blood contained:

	Urea.
Before nephrotomy	0.088 grammes.
Three hours and forty minutes later	0.093 ,,
Twenty-one hours later	0.252 ,,
Twenty-seven hours later	0.276 ,,

The urea increases, therefore, after the operation, and the increase takes place in a continuous manner proportional to the time.

The ligature of the ureters renders the kidneys totally inactive, for the blood which leaves this organ is found to contain the same quantity of urea as on entering. Hence, after the ligature of the ureters, following nephrotomy, urea accumulates in the blood.

The amount of urea excreted by man represents very nearly the whole amount of nitrogenous food which has failed to be assimilated, for the surplus is obviously found in the excrements, and they contain very little. The urine, therefore, is the liquid through which the nitrogen is eliminated, and the urea is almost the sole agent for effecting this.

For this reason the determination of the urea is highly important as furnishing us with data relative to the elimination of the nitrogen from the body.

Urea is not produced in the muscles; though creatin is easily transformed into urea when out of the body, yet, in spite of the considerable quantity of creatin which exists in the muscles, no urea is found in muscular tissue. On the contrary, it is sufficient to take in the food, creatin, gelatin, or analogous matters, to observe that the urea is thereby formed in greater quantity in the urine. It is therefore rational to admit that these substances are oxidized in the blood, and that their nitrogen is eliminated in the form of urea.

URIC ACID.—The urinary secretion furnishes each day 1.183 grammes of uric acid on an average (Wundt). It increases during digestion, and diminishes when the body is fatigued. In general it is produced whenever oxidation is impeded, and an increase in uric acid is associated with a corresponding diminution of urea. This acid is found in the urine of persons affected with the gout.

Uric acid, urate of ammonia, and urate of sodium are often deposited in urine a few hours after emission.

HIPPURIC ACID is found in small quantity in human urine. It increases with vegetable nourishment, in diabetes, and in certain other diseases. It is formed, molecule for molecule, when a benzoic compound is taken into the stomach. *Lactic acid* is only produced in the urine when digestion and respiration are impaired. It is formed in fevers, and whenever digestion and circulation are impeded.

Creatinin, and possibly *creatin*, exists in the urine.

An adult throws off, in the urinary secretion, about 1.16 gr. of creatinin in twenty-four hours. J. Munk (60-'76-1799) finds over ·008 per cent. sulphocyanhydric acid in normal urine.

Stoedler considers phenic acid and two ill-defined acids—*damolic* and *damaluric* acids,—to which the odour of the urine is supposed to be due, as constant constituents of the urine. Scherer regards *xanthin* as existing normally in the urine, though only in traces. It is an amorphous substance, soluble in acids and boiling water.

According to Schunck, urine contains always *indican*. This name is given to a body not as yet obtained in a crystalline condition, soluble in water, alcohol, and ether, and is essentially characterized by its property of decomposing in presence of strong hydrochloric acid, furnishing, by combining with water, indigo and a saccharine matter, indiglucin.

$$\underbrace{C_{26}H_{31}NO_{17}}_{\text{Indican}} + 2H_2O = \underbrace{C_8H_5NO}_{\text{Indigo}} + \underbrace{3C_6H_{10}O_5}_{\text{Indiglucin}}$$

The formation of this body accounts for the violet and reddish tints which are sometimes observed in urine undergoing decomposition. These phenomena take place only in the presence of atmospheric or other oxygen, as the indigo blue is very easily reduced.

$$2C_8H_5NO + H_2 = C_{16}H_{12}NO_2$$

Urozanthin and still more appropriately *indigogen* are modern synonyms for indican.

The substance which imparts to urine its yellow colour has been called *urochrome* by Thudicum. According to Heller, ether extracts from urine, which has been evaporated almost to dryness, a matter which he was not able to isolate, and which he calls *uroxanthin*. It is remarkable from the fact that, under the action of acids and in certain pathological states, it is transformed by oxidation into two other substances—one blue *uroglaucin*, the other red *urrhodin*.

Since these substances have not been isolated with certainty, we shall not further dwell on them.

GLUCOSE.—Glucose is always present in normal urine, according to some chemists, though doubted by Seegen and Gorup-Besanez.

The quantity of sugar present in normal urine amounts in twenty-four hours to 1 to 1.5 grammes according to Brücke, also according to Bence Jones. It is therefore less than one-thousandth.

FATTY BODIES, SALTS, AND GASES IN URINE.

Fatty bodies are found in the urine, but their proportion is very minute.

The quantity of saline matter in the urine is considerable. It amounts to about 15 grammes in twenty-four hours. This quantity may increase to 25 grammes, and decrease to 8 grammes. It is less in women, and still less in children. Among these solid matters are prominently *phosphates*, sodium phosphate, calcium phosphate, and magnesium phosphate. The quantity of phosphoric acid eliminated in the urine varies from 3 grammes to 5 grammes in twenty-four hours. This acid increases during digestion. It diminishes in pregnant women, and in the eighth month there is so little that both its reactions and those of calcium are hardly perceptible. Urine always contains alkaline chlorides, and chiefly sodium chloride. The quantity increases as the amount ingested increases, but the whole of this substance is not eliminated through the urine. The proportion of chloride increases after eating, and is at its minimum during the night. Exercise increases the amount. The weight of chlorine evacuated in twenty-four hours is about 10 grammes. When all salt is removed from the food the amount diminishes in the urine, and remains fixed at 2 to 3 grammes per day, which amount is derived from the tissues, and a rapid enfeeblement results. Sulphates are found in the urinary secretion. The quantity increases during digestion; it averages 2 grammes in twenty-four hours.

Normal acid urine contains no ammonium salts, but contains them on becoming alkaline, some time after its voidance. The same is the case with the urine of herbivora, which is always alkaline.

Many substances taken into the body which do not serve as aliments are found again in the urine, in case they are not capable of uniting with certain principles of the body to form insoluble compounds. Those metallic salts are among these latter, which form precipitates with albuminoid substances.

Substances not precipitable in the organism and difficultly oxidized—such as chlorides, iodides, sulphates, nitrates, urea, quinine, and most fragrant and colouring matters—reappear unchanged in the urine. Oxidizable substances, on the contrary, undergo the same transformations which they sustain when acted upon by oxidizing agents. Alkaline sulphides are converted into sulphates, alkaline organic salts into carbonates; benzoic and cinnamic acids into hippuric acid, uric acid into urea, salicine into saligenin and salicylic acid. The oxidation of certain other matters is more complete; they furnish carbon dioxide and water, which are the ultimate products of the oxidation of organic bodies. This is probably also what occurs to many substances which never reappear in the urinary secretion, even after abundant ingestion of the same;

such are mannite, ether, resins, the colouring matter of leaves, litmus, cochineal, amygdaline, anilin, camphor, etc.

The rapidity with which these bodies pass into the urine depends upon their solubility. Potassium iodide is found in the urine in a few minutes after being administered. A longer time is necessary for the urine to assume the odour which is developed after eating asparagus and the inhaling of the vapours of turpentine.

The gases of the urine are oxygen, nitrogen, and carbon dioxide. A mean of fifteen experiments made by Moring gave for a litre of fresh urine—

Oxygen	0.65 c.c.
Nitrogen	7.77 ,,
Carbon dioxide	15.96 ,,

These figures are probably too small, as the method by which the gases were determined was that of Magnus.

Walking increases the amount of carbon dioxide.

	Carbon dioxide.	Nitrogen.	Oxygen.
Urine during repose	11.877	7.494	0.493
,, when walking	22.880	8.204	0.466

The renal secretion of ophidians is solid, and composed chiefly of uric acid; that of batrachians is liquid, and contains urea.

The urine and excrements of birds contain chiefly acid urates, earthy phosphates, and a small amount of urea.

ANALYSES OF DIABETIC URINE.

PATHOLOGICAL STATES.—The urinary secretion increases in certain diseases (diabetes, polydipsia). In the first case its density may increase, as sugar is often present in large proportions; it sometimes is as high as 1.040. In polydipsia the density falls to 1.001. It diminishes in cholera, in diseases of the liver, and in fevers.

Diabetes.—The quantity of sugar excreted in the urine may amount to 1200 to 1500 grammes in 24 hours.

Bouchardat, to whom we are indebted for important investigations relative to this disease, has shown that the formation of sugar may be lessened or even arrested by submitting the patient to a nourishment devoid of farinaceous and saccharine matter, by furnishing him for example, instead of ordinary bread, bread made of *gluten* or flour freed from starch by washing.

The uric acid diminishes in quantity, or disappears in the urine of diabetic persons.

ANALYSES OF DIABETIC URINE BY SIMON AND BOUCHARDAT.

	Simon.		Bouchardat.
	I.	II.	I.
Density	1.018	1.016	..
Water	957.00	960.00	837.58
Solid constituents	43.00	40.00	162.42
Urea	traces.	7.99	8.27

	Simon.		Bouchardat.
	I.	II.	I.
Uric acid	traces.	traces.	traces.
Sugar	39.80	25.00	134.42
Alcoholic extract ⎫ Aqueous extract ⎬ Salts ⎭	2.10	6.50	5.27
Phosphates and mucus	0.52	0.80	0.24
Albumen	traces.	traces.	traces.
Oxide of iron	traces.	traces.	0.14

Markownikoff (72-182-362) finds acetone and ethyl alcohol, and believes they are formed from the glucose by fermentation.

Claude Bernard has shown that diabetes can be produced artificially by puncturing the "fourth ventricle."

A slow poisoning of frogs with curari, the slow action of strychnia, the destruction of the spinal column of frogs, etc., produce diabetes. Artificial diabetes is dependent upon the liver, as this state can never be obtained in a frog from which the liver has been removed. Saïkowsky has shown that if the formation of glycogenous matter in the liver of a rabbit be arrested, a result which is easily produced by the action of arsenates, this animal cannot become diabetic neither by curari nor by puncturing the fourth ventricle.

F. W. Pavy (112-23-59; 24-51) obtains diabetes

artificially in dogs by passing defibrinated arterial blood through the liver ; saliva used instead of blood produced no glycosuria. Upon inhalation of oxygen Pavy noticed a like appearance of sugar in the urine.

ALBUMINURIA.—Albumen does not exist normally in the urine. When it is found, it is due either to the secretion of an albuminous urine by the kidneys, or to an admixture of blood, pus, or lymph.

Albuminous urine is pale, acid, opaline, often of a density less than normal. As much as 20, 30 and even 35 grammes have been found to have been secreted in twenty-four hours.

The albumen increases after taking food ; it is at its minimum during the night. It increases with nitrogenous food.

According to Lehman this albumen exists in two states, one part is the modification of albumen called metaglobuline and paraglobuline, and is precipitable by carbon dioxide. The other remains in the liquid after the passage of the gas, and is precipitated by ordinary acids.

ANÆMIA.—The urine is pale and scarcely acid in anæmic persons; it sometimes even becomes alkaline. It is rich in salts and poor in most organic constituents.

OTHER ABNORMAL STATES OF THE URINE.—The urinary secretion decreases considerably in fevers, and is of a deeper colour and more dense than normal urine. Its acidity increases on account of the uric acid which forms abundantly, and of the lactic acid which is also developed. The urea disappears in about the inverse proportion. The extractive matters increase; the salts, and especially the sodium chloride, decrease.

The proportion of urea increases in intermittent fevers, also at the commencement of typhoid fever.

The quantity of urea, and especially that of uric acid, increases in inflammatory diseases. At the commencement of acute attacks the urea has been observed to amount to 60 grammes. The urine of persons affected with phthisis is richer in uric acid than normal urine, and fatty substances are also observed in it.

The urea diminishes in nervous affections.

In scarlatina and small-pox the urine contains ammonia, although it retains its acid reaction.

A. Pohl found cholesterin (40–76–737) in the urine of an epileptic patient who had taken large doses of potassium bromide.

Epithelial cells are found in large quantities in the urine in erysipelas, in scarlatina, in the commencement of Bright's disease, and in different urinary affections.

Fibrin and blood-globules appear in the urine during inflammation of the genital and urinary organs. In catarrh and in paralysis of the bladder the urinary secretion contains urate of ammonium. The urine is decomposed in the body of persons affected with catarrh of the bladder; and in the urine are observed monads, vibrions, and mycoderms.

Mucus is present in small quantity in normal urine. In various diseases of the genito-urinary organs, the mucus increases to such an extent as to render the urine turbid or milky.

Pus is found in the urine when suppuration is established in the genito-urinal tract.

The urine in jaundice contains the acids and colouring matters of the bile. These acids also pass into the urine in pneumonia. The bile itself is often found in the urine, and in this case boiling ether agitated with the urine takes on a green colour.

The urinary secretion diminishes or ceases entirely in cholera.

The proportion of phosphates increases in nervous affections. The quantity of chlorine decreases chiefly in pneumonia, in obstinate diarrhœa, and during cholera.

Chyle and casein are found in certain urines.

The urine is brown in acute rheumatism; it is red in many diseases in which the colouring matter of the blood passes into the urine; it is almost colourless in megrim and in nervous affections.

Von Merling and Musculus (60-1875-662) have examined the urine of a person who for a long time took 5 to 6 grammes of chloral hydrate every evening. The urine had an acid reaction, reduced alkaline copper solutions, contained neither chloroform, formic acid, nor sugar, but it contained chloral hydrate in small quantity, and turned the plane of polarization to the left; this latter property was due to an acid which they called *urochloral acid*, obtained by evaporating the urine acidified with sulphuric acid, and extracting the acid with a mixture of alcohol and ether. This new acid crystallizes in colourless silken needles, dissolves in water, alcohol, and a mixture of alcohol and ether, but is insoluble in pure ether; with potassium, sodium, barium, and copper it gives well crystallized salts; its composition is expressed by the formula $C_7H_{12}Cl_2O_6$.

F. Baumstark (60-1874-1170) found in the urine of a person suffering with leprosy two peculiar colouring principles which he calls *urorubrohematin* and *urofuchsohematin*. Urorubrohematin is a light bluish-black mass, insoluble in water, alcohol, ether, chloroform, or a solution of salt, soluble in alkalies, ammonium hydrate, alkaline phosphates and carbonates, alcohol containing acids, difficultly soluble in dilute sulphuric acid, and solutions of salt acidified with hydrochloric acid. The acid solution shows a characteristic absorption spectrum. The formula obtained by analysis is $C_{68}H_{94}N_8Fe_2O_{26}$ (?). Urofuchsohematin is black, pitchy, insoluble in water, alcohol, ether, chloroform, acids, or acidified or non-acidified salt solutions; it is soluble in alkalies, ammonium hydrate, alkaline phosphates and carbonates, and acidified alcohol. Analysis shows its formula to be $C_{68}H_{100}N_8O_{26}$ (?).

J. Müller (60-1874-1526) found in the urine of a child *pyrocatechin*.

URINARY SEDIMENTS.—Human urine abandoned to itself often deposits solid crystalline bodies. During fever, urate of sodium is observed to form a short time after emission. These crystals are microscopic, and the appearance of the deposit is corpuscular and colourless.

They are recognized by their disappearance when the urine is heated.

The urine sometimes deposits, three or four hours after emission, prismatic crystals of uric acid having a rhombic base.

When ammoniacal fermentation takes place in urine, a deposit of urate of ammonium is observed mingled with calcium phosphate or carbonate and ammonio-magnesium phosphate. This sediment forms whitish opaque grains, insoluble in water, soluble in acetic acid, and insoluble in ammonia.

At other times, crystals of calcium oxalate and ammonio-magnesium phosphate separate out.

C. Stein (1-187-99) finds in certain rare cases in which the urine is alkaline that magnesium phosphate occurs in the sediment.

There also separates out from the urine, under unusual and not well understood circumstances, an organic matter called *cystin*, containing sulphur.

This substance is colourless, insoluble in hot water, and soluble in ammonia.

Besides these crystalline substances, the urine deposits organized matters; mucus is alway present in it, sometimes pus, spermatozoids, blood globules, and coagulated albumen.

URINARY CALCULI.—This name is given to concretions of solid substances which form in the bladder. At times they escape with the urine in small grains or powder; they are then known as *gravel*.

These deposits are formed of various substances: uric acid, urate of sodium or ammonium, calcium carbonate, oxalate or phosphate, ammonio-magnesium phosphate, cystin or xanthic oxide.

The cystin may be obtained by treating the calculi with sodium carbonate and adding acetic acid to the liquid, when it deposits cystin in handsome hexagonal plates.

This substance may also be obtained from the kidneys.

A cystin calculus is soluble in caustic alkalies, and even in solutions of alkaline carbonates, with the exception of ammonium carbonate. It is dissolved by the mineral acids, and precipitated by acetic acid. Heated in the air, it furnishes sulphurous oxide. Heated with an alkali it furnishes a sulphide.

The nature of the calculi formed of cystin will be described further on.

ANALYSIS OF URINARY CALCULUS.

Urate of sodium	9.77
Calcium phosphate	34.74
Ammonio-magnesium phosphate	38.35
Calcium carbonate	3.14
Magnesium carbonate	2.55
Albumen	6.87
Water and loss	1.58
	100.00

(Lindbergson.)

ANALYSIS OF A FERRUGINOUS URINARY CALCULUS.

Ferric oxide	38.81
Alumina	23.00
Silica	17.25
Calcium	8.02
Water	10.89
Loss	2.03
	100.00

(Boussingault.)

ANALYSIS OF A CYSTIN CALCULUS.

Cystin	97.5
Calcium phosphate and oxalate	2.5
	100.0

(Lassaigne.)

ANALYSIS OF URINE.

THE whole of the urine voided during 24 hours is collected and its volume measured; of this 250 grammes are taken and allowed to stand for 24 hours; or the urine first voided in the morning after sleep is taken for analysis.

We commence by determining by means of litmus paper the reaction of this urine, and then determine its density; as the presence of water or albumen diminishes its density, while the presence of sugar and salts augments it. There are used for this test special areometers or hydrometers, called urinometers. It is well to verify once for all the graduation of these instruments by means of urines whose specific gravity has been determined by the ponderal method.

GLUCOSE.

We have already stated that abnormal urine may contain very large proportions of sugar: such urine is usually sweet and denser than ordinary urine. It is susceptible of fermentation, turns the plane of polarization to the right, and is but slightly coloured.

If it is desired to extract the sugar, basic lead acetate is added in excess, the solution filtered, the excess of lead precipitated by hydrogen sulphide, again filtered, and evaporated until it crystallizes.

THE QUALITATIVE TESTS.—Its presence merely may be detected by the tests given on page 187.

It should, however, be remarked that these reactions are not reliable unless a precipitate appears within one or two minutes boiling, as secondary reactions are produced with the other substances contained in the urine.

QUANTITATIVE DETERMINATION OF THE SUGAR BY THE REDUCTION OF COPPER SALTS.

PREPARATION OF THE LIQUID.—Weigh out 200 gr. of pure Rochelle salt, which place in a flask graduated to 1 litre; add 500 c.c. of a solution of sodium hydrate of 24° Baumé (D = 1.199), or 600 c.c. of a solution 22° Baumé (D = 1.180). The solution is facilitated by agitating and slightly heating in a water bath.

In another vessel dissolve 36.46 gr. of commercial copper sulphate, which has been purified by two or three recrystallizations, in 140 c.c. of distilled water, slightly heating. This solution is slowly poured into the first, stirring at the same time, that the precipitate may be dissolved. Rinse out the vessel which contained the copper sulphate two or three times, and after placing the litre-flask in a vessel of cold, common water, add enough distilled water to bring the liquid in the flask up to 1 litre. This solution is very reliable, and may be preserved for months exposed to the light without alteration. For an improved reagent, see p. 187.

Each 10 c.c corresponds to 0.050 gr. of pure cane sugar, or 0.0526 gr. of pure glucose.

The determination is made by placing 20 c.c of the cupro-alkaline solution in a porcelain dish, bringing the same to boiling, and adding gradually—at the same time agitating with a glass rod—the saccharine urine from a burette graduated to tenths of a cubic centimetre. There is first formed a yellowish, then a red precipitate. When the colour appears constant remove it from the flame; the supernatant liquid soon becomes clear; if it should appear greenish, again heat and add more of the urine drop by drop. The liquid must be neither greenish nor yellow. As long as there is any copper in the solution a drop of urine will produce an orange-coloured ring when it falls into the reagent. The amount of urine necessary to effect this will, of course, be an amount containing 2×0.0526 or 0.1052 gr. of glucose.

DETERMINATION OF GLUCOSE IN THE URINE, by means of lead acetate.—In clinical experiments it is often sufficient to add to the urine a few drops of a concentrated solution of lead acetate, separate the precipitate formed by filtering, and after bringing the filtrate to a known volume employ it in the same manner as the urine in the preceding operation.

ANALYSIS OF URINE—ALBUMEN.

The lead salt has the effect of precipitating the foreign matter. The glucose is not precipitated by the acetate unless ammonium hydrate is added.

When diabetic urine is highly charged with sugar it must be diluted with 5, 10, or 20 times its volume of water.

Glucose can also be determined by adding yeast to the urine, and from the loss of carbonic acid in the resulting fermentation calculating the glucose present. It can also be estimated by means of a polarizing apparatus, such as is used for determining the strength of saccharine solutions for sugar refineries.

As it is not within the scope of this work to supply elaborate instructions with regard to urine analysis, those desiring full details regarding the examination of urine for this or other constituents should consult some author on chemical analysis, or specifically on the chemical examination of the urine. A liberal amount of laboratory work is requisite, however, for such as would acquire a practical acquaintance with the chemistry of abnormal urine.

ALBUMEN.

Albumen is coagulated by heat and nitric acid. It is necessary to have recourse to these two reactions to detect with certainty the presence of albumen in urine. In fact, by simply heating the urine it often becomes turbid, owing to the precipitation of the earthy phosphates or carbonates; these salts may be recognized, however, by adding a drop or two of nitric acid, which will redissolve the precipitate formed. On the other hand, nitric acid will produce a white precipitate in the urine of a patient who has been taking various resinous remedies.

When it has been found that four to five cubic centimetres of urine coagulates on heating, and that it continues to coagulate after adding eight to ten drops of nitric acid, we may conclude that this urine contains albumen.

In order to estimate the amount of albumen we commence by ascertaining whether the urine is alkaline or not; in case it is, it should be slightly acidulated with acetic acid. 100 c.c. of the urine are taken and heated so as to cause coagulation—that is, until the urine just commences to boil. The liquid is then thrown upon a double filter, *i.e.*, two filters of equal size and weight placed one within the other. The albumen remains upon the inner filter; it is washed with water, then with alcohol, and when it has well drained the two filters are dried at 110°. The difference between the weight of the filters with the precipitate and the filters empty is the weight of the albumen.

Another determination to check the first may be made, precipitating the albumen with dilute nitric acid.

UREA.

We have already mentioned the importance of noting the variations in the amount of urea, since these variations give us light upon certain points in the process of nutrition. In order to ascertain whether a given urine is very rich in urea, a few drops are placed on a watch-glass with an equal volume of nitric acid and the glass floated on cold water; after a few minutes crystals of nitrate of urea are to be seen.

In order to determine the amount of urea, Leconte's method may be employed, which is based upon the oxidation of the urea by hypochlorites:—

$$CH_4N_2O + 3NaClO = 3NaCl + CO_2 + 2H_2O + N_2.$$

Carbon dioxide and nitrogen are disengaged: the former is absorbed by a solution of sodium hydrate, and the latter collected and measured; from the volume obtained the amount of urea can be determined.

BILE.

I. Gives with sub-acetate of lead **a greenish-yellow** precipitate.

II. Gives with a drop of nitric acid, green, blue, yellow, violet, and red coloration.

III. Gives with a solution of white of egg, on adding nitric acid, a precipitate which is **bluish-green**; **whereas** in the absence of bile it is **white**.

IV. Yields with tincture of iodine a green coloration.

According to W. G. Smith (7-[3]8-299) this reaction distinguishes bile from the so-called *indican*.

URIC ACID

Is recognized qualitatively by the test given on page 125. It is usually determined quantitatively by adding to a given amount of urine—not less than 150 to 200 c.cm.—sufficient hydrochloric acid to fully precipitate the uric acid, and allowing the liquid to stand for twenty-four to thirty-six hours. Traces of uric acid still remain in solution which, however, according to Neubauer, are compensated for by the amount of the urine pigment which also falls with the uric acid. The precipitate is filtered off, washed, dried, and weighed.

URATES.

The urates of sodium and ammonium are among the constituents of normal urine; they are often deposited after voidance when the urine has become cold; a deposit is then observed which disappears on slightly heating. These urates may be recognized by characteristics which will be given under Urinary Deposits.

HIPPURIC ACID.

If hippuric acid is found to exist in notable quantities in urine, it may be determined by the method already given under the general discussion of this acid.

CREATININ.

Creatinin may be detected and even quantitatively determined by the following method: Milk of lime, then calcium chloride, is added to 300 to 500 c.c. of urine until a precipitate no longer occurs; after being allowed to stand for a few hours the solution is filtered and the filtrate evaporated in a water-bath to the consistency of a syrup; 40 c.c. of 90 per cent. alcohol is then added, and the whole allowed to digest for twenty-four hours. The clear liquid is decanted off, and a solution of zinc chloride, as nearly neutral as possible, is added. A compound of zinc chloride and creatinin is formed, which is collected on a filter, washed with quite cold water, and dried.

INORGANIC SALTS.

The amount of salts in urine may be determined by evaporating 5 to 10 grammes in a porcelain dish. The residue is ignited at a slightly elevated temperature and weighed.

The chlorides, sulphates, phosphates, lime, etc., may be determined by the methods usually employed in inorganic quantitative analysis.

URINARY DEPOSITS.

If the urine has produced a deposit, its nature may be determined by plunging one end of a glass tube, which has been drawn out to a point, down into the deposit, the other end being closed by the finger; the finger is then removed, a quantity of the deposit allowed to run into the tube, the finger replaced, and the tube withdrawn. A certain quantity of the deposit is thus obtained, which may be tested with different reagents and examined under the microscope.

Urine which contains an excess of *uric acid* is acid and limpid; the deposit is then crystalline and slightly coloured, and is soluble in potassium or sodium hydrate, insoluble in ammonium hydrate or acetic acid. Nitric acid imparts a darker colour to urine rich in uric acid; a brown deposit may also be formed, which is soluble in alkalies.

Urine containing *urates* becomes turbid shortly after voidance; this deposit is white, or coloured and muddy. On heating it dissolves, as well as by adding potassium or sodium hydrate. Sometimes this deposit is coloured.

Urine containing *earthy phosphates* may become turbid, but this deposit cannot be confounded with the preceding, as it does not dissolve on heating, is soluble in acetic acid, while not soluble in potassium or sodium hydrate.

Urinary deposits formed of *calcium oxalate* are **white**;

they are insoluble in **ammonium hydrate and acetic acid**; they also do not dissolve on heating, but are soluble in mineral acids. If the deposit were formed of *calcium carbonate*, it would dissolve in acetic acid with the disengagement of carbon dioxide. Deposits of *ammonio-magnesium phosphat* are white; soluble in acetic acid, insoluble in ammonium hydrate.

Urine containing *cystin* has an acrid and even repulsive odour. It furnishes a deposit which does not dissolve on heating, and is soluble in ammonium hydrate.

Certain urines become turbid on account of the *mucus* they contain, or because decomposition has set in. The presence of *blood* renders the urine red, the presence of *bile* greenish. Urines are sometimes met with which are whitish or opalescent; agitation with ether renders them clear. Blue and blackish urines also occur.

If a drop or two of a urinary deposit is viewed *through a microscope* **magnifying** 250 diameters, and the preceding reactions employed, they will appear much more distinct. We would, however, add the following :

Uric acid occurs in crystalline plates of a diamond shape : their angles are often rounded off. These plates are often isolated, sometimes united in the form of rosettes and stars, and rarely in the form of needles.

The urates are sometimes amorphous, sometimes crystalline. Deposits of urates may be distinguished from those of uric acid by their solubility in hot

K

water. They are generally found when the urine is alkaline.

Crystals of urates, heated with a small quantity of nitric acid, give a residue of uric acid. More nitric acid forms alloxan, as do deposits of uric acid, and this yields a characteristic red colour with ammonium hydrate.

Calcium phosphate is amorphous.

Ammonio-magnesium phosphate occurs in prismatic crystals.

Calcium oxalate crystallizes in regular octahedrons.

Cystin, $C_3H_7NSO_2$, occurs in beautiful hexagonal plates. It is obtained by treating the deposit with ammonium hydrate, and allowing the liquid to stand; the cystin separates out, and by the aid of the microscope the form of the crystals may be distinctly seen. Under these conditions the uric acid would not dissolve, a fact which permits of distinguishing between deposits of cystin and those of uric acid. Cystin is neutral, insoluble in water, alcohol, ether, or acetic acid. It is soluble in the mineral acids, also in oxalic acid. Ignited on platinum foil, it gives off an alliaceous odour. It is coloured, like uric acid, upon treatment with nitric acid and ammonium hydrate. It dissolves in alkaline solutions. Heated with potassium or sodium hydrate in presence of lead oxide, it blackens on account of the formation of lead sulphide. Cystin is of rare occurrence, and its physiological and chemical relations have not been fully studied. Loebisch (1 182-231) has shown that no diminution of urea or uric acid occurs in cases of cistinuria, though earlier investigators, and recently also Nieman (1-187-101), have come to the conclusion that uric acid at least decreases. Nieman established in the same research that there is no change in amount of sulphur in urine by reason of the presence of cystin.

Pus may be recognized by the spherical globules, in which two or three nuclei are observed, on the addition of acetic acid. This matter is converted into a jelly-like mass in contact with potassium or sodium hydrate.

Mucus may be distinguished by its ropy consistency and its coagulation with acetic acid; various kinds of cells are observed floating in the liquid. In these deposits epithelium cells are almost always found; they are oval or irregular.

We also find in urinary deposits:

Blood Globules.—If the urine remains acid, they appear as quite characteristic discs; if the urine becomes alkaline, they are destroyed.

Tube Casts.—These may be: *epithelial, fibrinous, mucous hyalin,* (or colloid) and *amyloid.*

The first have special diagnostic importance in diseases of the kidneys. These casts are generally nearly straight, though sometimes curvilinear, and not unfrequently are difficult to find. The epithelial cells which cover them are nearly normal in appearance.

Epithelial Cells.—These may originate from the kidney, the bladder, the ureters, or the canal of the urethra.

Vibrions.—Linear in form, and exhibiting characteristic movements.

URINARY CALCULI.

PHYSICAL ASPECT.—1. *Uric Acid.*—Form, round; colour, brown or reddish; fracture, earthy or partially crystalline. When sawn through, a powder is obtained resembling the sawdust of wood.

2. *Urate of Ammonium.*—These calculi are small, and of a clay or ash colour, with an earthy fracture. They are formed in concentric layers.

3. *Cystin.* — These calculi are voluminous, pale yellow, rounded in form, glossy, crystalline, and sometimes striated.

4. *Calcium Oxalate.*—Calculi of this substance are called *mulberry* calculi, from their resemblance to the fruit of the mulberry-tree, their surface being covered with rounded tubercles. They are usually grey, though sometimes dark brown, which colour is due to the organic matter which covers them. Their fracture usually is granular, sometimes crystalline.

5. *Ammonio-magnesium Phosphate.*—These calculi are white, crystalline, semi-transparent, covered with small brilliant crystals; they are very easily pulverized.

6. *Calcium Phosphate.*—These calculi are white, amorphous, and formed in concentric layers.

The following table indicates in brief the method to be followed in examining different calculi. We should mention, however, that calculi are not always composed of a single substance; they are quite frequently formed of several compounds. This table of reactions applies as well to urinary deposits.

CHEMICAL EXAMINATION. 369

CHEMICAL EXAMINATION.—A small fragment of the calculus is reduced to powder, and several successive portions are subjected to the following tests:—

Heat a small portion on platinum foil, lightly and slowly.

- Volatilizes completely.
  - Ammonia is not disengaged when it is heated with potassium or sodium hydrate.
    - It dissolves in nitric acid; the evaporated solution becomes purple when exposed to vapour of ammonia. — **Uric acid.**
  - Ammonia is disengaged when it is heated with potassium or sodium hydrate.
    - It presents the same reactions with nitric acid and ammonia. — **Ammonium urate.**

- Burns, emitting an alliaceous odour . . . — **Cystin.**

- A residue remains; this is moistened with a few drops of water.
  - The liquid is alkaline
    - It dissolves in potassium or sodium carbonate. — **Xanthin.**
    - It does not dissolve . . . — **Calcium oxalate.**
  - It is not dissolved. . . . . — **Silica.**
  - The liquid is not alkaline; an excess of nitric or hydrochloric acid is added.
    - It dissolves; a portion of the original powder is heated with an excess of potassium or sodium hydrate.
      - Ammonia is disengaged. — **Ammonio-magnesium phosphate.**
      - Ammonia is not disengaged. — **Calcium phosphate.**

CUTANEOUS SECRETIONS OR TRANSPIRATIONS.

We include under this head the products of the sebaceous follicles, of the glands of Meibomus, and the wax of the ears.

These contain an albuminoid substance, of which but little is known, neutral fatty bodies (stearin, olein), epidermic cells, and epithelium and other cells, sodium chloride, ammonium chloride, and alkaline and earthy phosphates.

SWEAT.

The quantity of this secretion has not yet been determined. It is, however, known that it is quite large, and it is believed to be more than half of that of the pulmonary exhalations.

It is obtained by pressing sponges against the skin while in perspiration, and afterwards washing these sponges with water.

Sweat is an acid liquid, of an odour variable with individuals, and of a saline taste. It leaves 1 to 2.5 per cent. of fixed substances on evaporation at 100°. Sodium chloride, mixed with potassium chloride, forms two-thirds of this residue. Alkaline phosphates have not been found in it. Its acidity is due to acids of the fatty series: the most abundant is formic acid associated with small quantities of acetic and lactic acids. Favre has detected in it the existence of a special acid—sudoric acid.

Sweat contains fatty matters derived from the

sudorific and sebaceous glands and a nitrogenous substance (possibly urea), which readily changes into ammoniacal salts. In uræmia, the perspiration of the face contains a considerable quantity of this substance.

The sweat appears milky, on account of the epithelial cells with which it is charged. It contains nitrogen and carbon dioxide gases.

THE SPERMATIC FLUID, OR SEMEN,

Is viscid, opaque, heavier than water, and possesses a marked odour. Heat does not coagulate it. It is precipitated by alcohol and acids.

It is formed of a colourless fluid, in which float a large number of very minute bodies, called spermatozoids. In man they have a flattened or oval body, to which is joined a long filiform "tail."

The movements are principally executed by the tail, which has a sort of vibratile undulatory motion.

The seminal liquid gelatinizes after emission. This effect is attributed to an albuminoid matter called *spermatin*, which is a substance resembling globulin and mucin. Heat does not coagulate its solutions. Acetic acid renders them turbid, and an excess of the acid re-dissolves the precipitate. These solutions are precipitated by potassium ferrocyanide and nitric acid.

After having been evaporated to dryness, this substance no longer dissolves in water, but is dissolved in very dilute alkaline solutions.

The fecundating property of the spermatic fluid rests in the spermatozoids. They preserve vitality for a long time in the urine, and even in a dry state. If a cloth impregnated with dry sperm be moistened and placed on the stage of a microscope, the active spermatozoids are readily perceived. Spots of semen heated slightly for a few minutes assume a dark yellow colour.

The seminal liquid contains in suspension, besides the spermatozoids, white granular corpuscles, mucus, and *débris* of epithelium. It holds in solution, in addition to spermatin, lecithin, various fatty bodies, sodium carbonate—which renders it alkaline—sodium chloride, and phosphates.

MUCUS FLUIDS OF THE SEROUS MEMBRANES.

Mucus is a viscous, ropy liquid, containing epithelial cells and small colourless corpuscles, few in number in a normal state, but which increase greatly when the membranes are inflamed. The composition of mucus in different parts of the body presents differences not yet determined.

Mucin is the name given to that principle of which, however, little is known, imparting to mucus its ropy consistency. It is found in a number of the fluids of the body. Eichwald has given a process by means of which he extracts this substance from different liquids or tissues.

It is most readily extracted from pulmonary expectorations.

These are diluted with water, and an excess of acetic acid added. The turbid liquid is washed on a filter with dilute acetic acid as long as the filtrate gives a precipitate with potassium ferrocyanide. The solutions are then treated with lime water and the mucin precipitated from the solution by acetic acid. This body appears to be largely soluble in water; it is precipitable by alcohol and dilute acids, and soluble in alkalies.

It is distinguished from albumen in not coagulating by heat. It also furnishes tyrosin under the action of dilute sulphuric acid.

G. Gaelchli (18-78-77) found that mucin on putrefying generated indol, phenol, and a sugar-like substance.

Normal mucus does not contain albumen.

An analysis of nasal mucus by Nasse yielded:

Water	933.7
Mucin	53.3
Lactates and extract soluble in alcohol	3.0
Extract soluble in water and phosphates	3.5
Alkaline chlorides	5.6
Sodium hydrate	0.9
	1000.0

Urine left standing for a short time often deposits mucus which is whitish, soluble in the alkalies, and partially in acids. It facilitates the transformation of the urea present into ammonium carbonate.

SEROSITY effects the lubrication of various surfaces of the body, preventing friction; its composition varies slightly in different organs. Albumen, mucus, and soda are ordinarily found in it.

Synovia is the serosity which lubricates the joints. It is dense and slightly alkaline. It differs from mucus in containing albumen.

According to Berzelius, it contains:—

Water	926
Albumen	64
Extractive matters and soluble salts	6
Calcium phosphate	1.5

Its composition, however, varies according to amount of exercise taken.

ANALYSIS OF THE HYDROCEPHALUS FLUID.

Mucus with a trace of albumen	0.112
Sodium carbonate	0.124
Sodium chloride	0.664
Potassium chloride and sulphate	traces
Calcium phosphate	,,
Magnesium phosphate	,,
Iron phosphate	0.020
Water	99.080
	100.000
	(Marcet.)

ANALYSIS OF THE HYDROPSICAL FLUID.

Albumen	2.38
Urea	0.42
Sodium chloride	0.81
Sodium carbonate	0.21
Sodium phosphate, with traces of sodium sulphate	0.06
Mucous substance	0,89
Water	95.23
	100.00

(Marchand.)

VESICULAR SEROSITY.

Coagulable albumen	5.25
Albumen more soluble in water	0.50
Salts	0.26
Water	93.99
	100.00

(Brandes and Reimann.)

COLLOIDIN. — Gautier, Cazeneuve, and Daremberg (97-[2] 21-482) have examined the jelly-like contents of a large ovarian cyst: they diluted the same with water, heated to 110 degrees in closed vessels, filtered after allowing to cool, dialyzed the filtrate in order to remove the salts, and precipitated with alcohol, whereby they obtained a white flocculent mass, soluble in water, and not precipitated either by metallic salts or mineral acids, but precipitable by tannic acid and alcohol. They have called this substance colloïdin, and give as its formula $C_9H_{15}NO_6$. According to Gorup-Besanez this body is closely allied to mucin.

MILK.

Milk is a white, opaque liquid, inodorous while cold, and of a slightly sweetish taste.

Its density varies but little:—

Human milk	1.0320
Cows' ,,	1.0300
Goats' ,,	1.0341
Asses' ,,	1.0355
Sheep's ,,	1.0409

Human milk is alkaline. The milk of herbivora has generally the same reaction. That of carnivora is believed to be acid; at least it acidifies so quickly when once drawn that it is difficult to state its reaction positively.

Milk is formed of an almost colourless and transparent liquid, in which float an immense number of oleaginous globules. These globules are visible only under the microscope; their size varies from 0.0027 m.m. to 0.0041 m.m. They are opaque, and it is to these globules that the opacity of the milk is due. The fatty bodies of which they are formed are probably contained in an **albuminoid membrane**. If to milk we add a little potassium hydrate and ether, the alkali dissolves the membrane, the ether absorbs the fatty bodies, and the milk is changed into a limpid, transparent liquid. On placing some milk under a microscope, and moistening it with a drop of acetic acid, the membrane will be seen to be attacked, and the fatty bodies will immediately run together, while if it be simply agitated with ether, the globules remain unchanged.

Robin, however, supposes that the milk globules have no special envelope, but are surrounded by a thin layer of a saponaceous matter formed of fatty bodies, salts, and albuminoid compounds.

Milk left to itself separates into two layers; that formed above, by the union of the globules, constitutes the *cream*, that below forms a white liquid, having a slightly blue tinge.

On subjecting milk to a violent and prolonged beating, the globules unite and separate from the liquid, and butter is obtained. The fatty bodies of milk are formed of several principles:—

Butyrin, caproin, caprin; about . .	2.
Olein	30.
Margarin	68.

And a small amount of stearin.

But these proportions are necessarily very variable.

E. Tisserand (46–[3] 9–440) has summarized the following data:—

I. The separation of cream occurs the more promptly according as the temperature approaches 0°.

II. The lower the temperature the larger the volume of cream and the yield of butter; at the same time the butter milk, butter, and cheese, are all of a better quality.

In human milk the mean percentage of butter is 2.42. It ranges between 2.80 and 3.50 in cows' milk.

According to different experimenters the margarin is very impure; it contains stearin, myristin, and even other compounds.

The lower layer contains various substances, of which the principal ones are :—

Casein, an albuminoid matter previously described: the milk contains more of this substance after a nourishment of nitrogenous food than after one of vegetable matters.

Sugar of milk.

Different salts, principally phosphates and chiefly calcium phosphate; sulphates are not present.

Milk allowed to stand in the air rapidly loses its alkaline reaction and becomes acid. It then coagulates. This effect is due to the lactic acid which forms spontaneously in milk. It is formed by a fermentation called *lactic* fermentation.

The sugar of milk is the substance which is transformed into lactic acid with the co-operation of nitrogenous ferments.

The coagulum is formed of casein and fatty substances the liquid which remains is known as *butter milk*.

A. Vogel (75-23-505) confirms the observations of Schwalbe (36-1872-833) that oil of mustard prevents the coagulation of milk; according to his investigations the formation of lactic acid is in a great measure hindered by the presence of the oil of mustard. Oil of bitter almonds and oil of cinnamon prevent the formation of this acid to a less degree, while oil of turpentine, oil of cloves, benzol, carbolic acid, carbon bisulphide, and hydrogen sulphide are almost without action.

It is an alkali, soda, which holds the casein in solution in fresh milk, and milk may be kept fresh for a very long time by simply adding to it a few thousandths of an alkaline bicarbonate. On the other hand, milk will at once coagulate on the addition of an acid.

Besides the acids, a large number of substances possess the property of causing milk to coagulate; such are alcohol, tannin, different salts, many plants which are not acid, the flowers of the artichoke, of the thistle, and of the butter wort (*Pinguicula vulgaris*), which render it ropy, and especially rennet, a substance obtained from the stomachs of sucking calves. One part of rennet will coagulate 30,000 parts of milk, and the wooden vessels which have contained rennet, and which are used in dairies, may be used for a very long time for the operation without any subsequent addition of this substance. According to certain experimenters rennet effects the transformation of a certain amount of the sugar of milk into acetic acid; according to others this transformation is produced by an albuminoid substance called *chymosin*. The coagulum of milk is employed in making cheese.

The nature of the food influences the character and quantity of this secretion. The butter increases if the food contains much fatty matter and when the food is vegetable. A mixed or animal diet diminishes the proportion of butter, and increases the proportion of casein and sugar.

Fasting diminishes the secretion. The milk is then poor in sugar and salts, and becomes rich in fat and casein.

During certain affections of the mammillary glands, mucus, infusoria, fibrin, and epithelial *débris* are found in the milk.

Albumen occurs in the milk when the mammillary glands are the seat of inflammation. In Bright's disease urea passes into the milk.

COMPOSITION OF MILK, BY BOUSSINGAULT.

	Human.	Cow.	Ass.	Goat.	Mare.	Dog.
Water	88.4	87.4	90.5	82.0	89.63	66.30
Butter	2.5	4.0	1.4	4.5	traces	14.75
Sugar of milk and soluble salts	4.8	5.0	6.4	4.5	8.75	2.95
Casein, albumen, and insoluble salts	3.8	3.6	1.7	9.0	1.60	16.00
	99.5	100.0	100.0	100.0	99.98	100.00

Mott (100-6-364) finds milk of the negro race richer in solid matter than that of the Caucasian.

According to recent investigations of Lieberman (1-181-102) there is another albuminoid substance in milk besides those given in the foregoing table, but which has not yet been isolated.

COMPOSITION OF THE MILK OF A WOMAN, AT DIFFERENT PERIODS, BY SIMON.

Days after Childbirth.	Specific Gravity.	Water.	Dry Residue.	Casein.	Sugar.	Butter.	Mineral Salts.
2	1.0320	82.80	17.20	4.00	7.00	5.00	0.316
10	1.0316	87.32	12.68	2.12	6.24	3.46	1.180
17	1.0300	88.38	11.62	1.96	5.76	3.14	0.166
18	1.0300	89.90	10.10	2.57	5.23	1.80	0.200
24	1.0300	88.36	11.64	2.20	5.20	2.64	0.178
67	1.0340	89.32	10.68	4.30	4.50	1.40	0.274
74	1.0320	88.60	11.40	4.52	3.92	2.74	0.287
82	1.0345	91.40	8.60	3.55	3.95	0.80	0.240
89	1.0330	88.06	11.94	3.70	4.54	3.40	0.250
96	1.0334	96.04	10.96	3.85	4.75	1.90	0.270
102	1.0320	90.20	9.80	3.90	4.90	0.80	0.208
109	1.0330	89 00	11.10	4.15	4.30	2.20	0.276
117	1.0344	89.10	10.90	4.20	4.40	2.00	0.268
132	1.0340	86.14	13.86	3.10	5.20	5.40	0.235
136	1.0320	87.36	12.64	4.00	4.00	3.70	0.270

According to Berzelius, skimmed cows' milk contains:—

Casein, with a small quantity of butter	2.600
Sugar of milk	3.500
Alcoholic extract, lactic acid, lactates	0.600
Potassium chloride	0.170

Alkaline phosphate 0.025
Calcium phosphate, lime combined with casein, magnesia, and traces of iron oxide 0.230
Water 92.875

100.000

H. Ritthausen (18-'77-348) has recently found in milk another carbohydrate, differing from milk sugar, and more resembling dextrin.

FLESH.

We can have only imperfect ideas in regard to the transformations which the plastic principles (albumen, fibrin, casein) undergo in being converted into assimilable matter and tissue, also as to the manner in which each organ selects from the nutritive substances the elements which are suited for its use.

It is certain that albumen plays the principal rôle, for it is observed to give rise to fibrin and other nitrogenous substances under certain circumstances, and especially in the incubation of the egg; certain physiologists have also thought that in digestion all nitrogenous substances are converted into albumen, and that in nutrition the albumen is changed into fibrin, a substance which, from the facility with which it coagulates, is the principal agent in the creation and renewal of the tissues, that is, of the solid portions of our bodies.

These ideas are probably exaggerated, or at the least their correctness has not been demonstrated.

MUSCULAR TISSUE.—The muscles are constituted of a reddish contractile tissue, formed of fusiform elongated cells and of striated filaments, constituting an external envelope, called the sarcolemma, and of internal substances, from which a variety of fibrin, syntonin, may be extracted.

This latter is probably the substance into which albuminoid matters are changed during digestion in the stomach (parapeptone).

Solutions of syntonin in acids are not coagulated by boiling; they are precipitated by chlorides and alkaline sulphates. Syntonin dissolves in caustic alkaline liquids and in dilute solutions of carbonates, and is reprecipitated when these solutions are neutralized, even in the presence of alkaline phosphates. This last character distinguishes it from the albuminates.

The fibres of the muscles are surrounded by a fluid which may be considered as the plasma of the muscles. It may be prepared according to Kühne by removing the muscles of an animal recently killed, and freezing them at a temperature of about $-7°$, whereby they become very brittle. They are pulverized in a well-cooled mortar, and thereupon subjected to a heavy pressure in an appropriate press. A liquid is thus obtained, which is placed upon a filter surrounded by a refrigerating mixture. The liquid, which filters very slowly, is opaline-yellowish, viscid, and alkaline. It coagulates at ordinary temperatures, furnishing *myosin*, which may be readily obtained by causing the filtrate to fall into water at the ordinary temperature. If acid solutions of myosin are saturated, it is then no longer this body which precipitates but *syntonin*. Syntonin differs from myosin by not dissolving in solutions containing less than 10 to 12 per cent. of common salt. Myosin may also be obtained more simply by pounding flesh with water containing 8 to 9 per cent. of common salt. After allowing this to stand twenty-four hours it is filtered by being pressed through cloth, and the myosin precipitates on pouring the liquid into water.

The liquid which remains after the coagulation of myosin contains, according to Kühne, two albuminoid substances, one coagulable at 75°, the other at 45°, and alkaline albuminates; also salts, which are chiefly phosphates, lactic acid, and lactates; sugar and various organic substances, as creatin, creatinin, inosic acid, inosite, sarcosin, sarkin, and xanthin. This liquid is coagulable by heat, and of a red colour; its acidity is due to lactic acid and acid phosphate of potassium, which may be extracted from the muscles by dilute alcohol.

It is claimed by Frémy and others that there exists in the muscles a special acid, called *oleophosphoric acid*, and that this acid is combined with sodium.

According to Dubois Reymond, the muscles do not possess an acid reaction until after death, and while contractile their reaction is slightly alkaline.

In certain pathological states, urea, uric acid, and various other products are present.

H. Struve (60-'76-623) finds in fatty muscular tissue a new body which gives the absorption spectra of blood, but is changed, unlike the latter, by the action of alkaline sulphides and acids.

COMPOSITION OF FLESH.

	Pectoral Muscles.	
	Man.	Woman.
Water	72.46	74.45
Muscular fibres, vessels, and nerves	16.83	15.54
Fats	4.24	2.30
Extractive matters	2.80	3.71
Cellular tissue	1.92	2.07
Soluble albumen	1.75	1.93
	100.00	100.00

(Von Bibra.)

Flesh leaves from 2 to 8 per cent. of ash, formed chiefly of alkaline and earthy phosphates; sodium chloride and sodium sulphate are also present.

The broth produced by digesting muscular tissue in water contains, according to Chevreul:—

Water	988.57
Organic substances dried in a vacuum	12.70
Salts (phosphates, sulphates and chlorides of potassium, sodium, calcium, magnesium, and iron)	3.23
	1004.50

CREATIN, $C_4H_9N_3O_2 + H_2O$.
CREATININ, $C_4H_7N_3O$.
SARCOSIN, $C_3H_7NO_2$.

These three bodies have the highest importance in connection with the study of muscular tissue.

Liebig found in:

Muscular flesh of the ox . 0.69 creatin.
„ „ „ horse 0.72 „

Creatin is prepared by treating meat cut into very small pieces with cold water as long as anything is dissolved, and the solution evaporated; the concentrated liquid, filtered, furnishes creatin.

This body occurs in rectangular prisms, without taste or odour, soluble in 74 parts of cold water, but more soluble in boiling water.

On being boiled with strong acids it furnishes creatinin.

$$HCl + C_4H_9N_3O_2 = H_2O + \underbrace{C_4H_7N_3O.HCl}_{\text{Chlorhydrate of creatinin.}}$$

This chlorhydrate decomposed furnishes creatinin as a crystalline alkaline substance, more soluble than creatin in both water and alcohol. Creatinin may be reconverted into creatin by boiling with lead oxide.

It forms with zinc chloride a combination which is but slightly soluble in cold water. According to Neubauer, creatin does not exist in flesh, but creatinin only, and the creatin which is found is formed by the transformation of the creatinin. Creatinin also exists in urine, and in the muscles of the crustacea.

On submitting creatinin to a prolonged ebullition with baryta water another substance is formed, called *sarcosin*.

$$H_2O + C_4H_9N_3O_2 = \underbrace{CH_4N_2O}_{\text{Urea.}} + \underbrace{C_3H_7NO_2}_{\text{Sarcosin.}}$$

This body crystallizes in rhombic crystals, which are colourless, very soluble in water, somewhat soluble in alcohol and insoluble in ether. Sarcosin melts at a temperature above 100°, and is volatile. It possesses the characters of glycocol and its homologous substances.

INOSIC ACID.—The mother liquor of creatin is acid, and has an odour of meat broth. Extract of meat treated with baryta furnishes on evaporation inosate of barium, and the liquid contains inosite. The formula of inosic acid is usually given as $C_5H_8N_2O_6$, though some authors regard it as $C_{10}H_{14}N_4O_{11}$ (21-269).

XANTHIN.—$C_5H_4N_4O_2$. To prepare this substance muscular tissue is well beaten, and alcohol and water successively added as long as anything is dissolved. These two liquids are now united and heated, in order to coagulate the albumen and drive off the alcohol. The liquid is filtered, and lead subacetate added. The precipitate is collected, washed, and decomposed with hydrogen sulphide while suspended in water. The filtered liquid is boiled and evaporated. The xanthin deposits in a non-crystalline mass. It can also be prepared from the liver. Xanthin is soluble in cold water, alcohol, and ether. It forms with acids salts which are generally crystallizable; it is precipitated even from very dilute solutions by mercuric chloride or nitrate.

It dissolves in alkaline liquids. If calcium hypochlorite be added to one of these solutions, a greenish precipitate is formed which becomes brown and then disappears. This reaction is quite delicate, and a useful test.

OTHER TISSUES.

CELLS.—The cells are the simplest structures of the body. Their mass is very minute, and their form variable. They are not always enclosed by an envelope, and they vary in their chemical nature. They contain one or more nuclei, and when new a gelatinous liquid (*protoplasm*), capable of contractile movements under the influence of chemical agents, of electrical or mechanical forces. If old, they contain different matters, derived either from modifications of the protoplasm or the introduction of foreign substances.

The protoplasm coagulates after death. It appears to contain myosin, also other albuminoid, fatty, and saline constituents.

AREOLAR TISSUE is chemically characterized by the action which hot water has upon it.

At first it swells, assumes a jelly-like appearance, and finally dissolves, producing gelatin, which, on cooling, is of a tremulous consistency.

Dilute inorganic acids and dilute alkalies also effect this transformation. There is believed to exist in this tissue a substance (collagene, glutine, geline) analogous with ossein, which, in contact with hot water, furnishes gelatin; also a substance (elastin) not furnishing gelatin.

Tannin and mercury dichloride form with these matters imputrescible compounds.

Cellular tissue is converted into a transparent and colourless jelly by the action of strong acetic acid; but the fibre is not attacked, for if the acid be saturated with ammonia water it reappears in its ordinary condition.

The *elastic tissues* do not dissolve even after an ebullition of sixty hours, and do not furnish gelatin.

Hydrochloric acid dissolves them, turning brown at the same time. With sulphuric acid they furnish leucin and not gelatin. This may be obtained quite pure by boiling cellular tissue with water, then with acetic acid, and macerating the residue with a dilute alkaline solution. To the product thus obtained the name of *elastin* has been given.

The *mucous areolar tissue* differs chemically from ordinary conjunctive tissue, in that it does not furnish gelatin on being boiled with water.

The *reticular tissue* of the cutis contains the pigment called *melanin*, the colouring matter of the skin. This tissue is not reproduced completely where destroyed, but is replaced by cellular tissue, and the cicatrix is due to the fact that this latter tissue is colourless.

The *epidermis* furnishes gelatin on boiling with water. It appears to contain iron, and H. P. Floyd (84-34-179) has found it to contain in the negro twice as much of this element as in whites.

Sulphuric acid softens and dissolves it, nitric acid colours it yellow, alkalies dissolve it, the sulphides render it of a brown colour, and silver salts blacken it.

The epidermis, hair, bristles, feathers, nails, horns, and epithelium have an almost identical composition.

	Epidermis.	Epithelium.	Hair and Bristles.	Nails.	Feathers.	Horn.
Carbon	50.34	51.53	50.00	51.00	52.42	50.94
Hydrogen	6.81	7.03	6.40	6.82	7.21	6.65
Nitrogen	17.22	16.64	17.00	17.00	17.89	16.28
Oxygen and sulphur	25.63	24.80	26.60	25.18	22.48	26.13
	100.00	100.00	100.00	100.00	100.00	100.00

The horny tissues are formed of cells containing nuclei which have united and dried. Indeed, when these different tissues are treated with alkaline solutions, ovoid cells are seen, each containing a nucleus. Sulphuric acid likewise renders this structure apparent. This tissue leaves about 1 to 1.5 per cent. of ash on ignition.

Horn treated with fused potassium hydrate and with dilute sulphuric acid, furnishes tyrosin and leucin.

Hydrochloric acid renders it blue, nitric acid yellow; aqua regia attacks it with energy.

Feathers possess the same general properties. The colour of the feathers is due to different pigments, rarely soluble in water, sometimes in ammonia, and ordinarily in alcohol. They generally contain less oxygen and more silica than horn and analogous tissues.

Hair has the same composition and chemical characters as horny tissue.

Its colour is due to oils of various tints. With age this oily secretion ceases to be produced and the hair whitens; the white colour seeming to be due to the fact that the tubes contain no secretion, but are filled with air. The fatty bodies of the hair are formed from the volatile acids of perspiration, and also of margarin, olein, and stearic acid. Hodgkinson and Sorby obtained (28-222-592) from black hair and feathers a black pigment, to which they ascribe the formula $C_{18}H_{16}N_2O_8$.

Hair contains 0.5 to 2.0 per cent. of inorganic substances, containing a considerable proportion of iron and small quantities of silica. Mulder found in epidermis an organic sulphur compound he called *keratin*.

CARTILAGINOUS TISSUE.—The cartilages are ordinarily formed of a flexible tissue, whose composition is not greatly different as to its organic constituents from that of the preceding substances, though varying in organic composition with age and in the different parts of the body :—

Carbon	50.91
Oxygen	6.96
Nitrogen	14.90
Oxygen	27.23
	100.00

Hoppe-Seyler found in a proximate analysis of cartilage from the knee of a man aged twenty-two years:—

H_2O	75.59
Organic matter	24.87
Inorganic matter	1.54
	100.00

This tissue is not homogeneous; under the microscope it appears composed of a colourless fibre and cells containing granulated protoplasm. The matter of the cells is different from the gelatinous substance forming their envelopes. It does not dissolve in boiling water even under pressure.

The cartilaginous substance proper, *cartilagein*, furnishes, with boiling water, a substance which resembles gelatin in its composition, but from which it differs in several characteristics, and especially by its giving a precipitate with acids, lead acetate, and alum, while gelatin gives no reaction with these substances. It is called by the name of *chondrin*. Chondrin turns the plane of polarization to the left. Treated with hydrochloric acid it furnishes a variety of glucose (*chondroglucose*) and various nitrogenous substances of which little is known.

A distinction has lately been made between the cartilages just spoken of and the fibro-cartilages. These last contain a fibrous matter without nuclei, differing from the ordinary cartilaginous substance by producing with boiling water a substance which is but slightly precipitated by tannin. The fibro-cartilage of the knee must also be distinguished from the preceding, from the fact that it produces gelatin with boiling water.

Cartilaginous tissue contains 55 to 75 per cent. of water, 2 to 5 per cent. of fatty bodies, and 1.5 to 6 per cent. of mineral substances.

NERVE TISSUE.—Of it are composed the nerves, ganglia, brain, and the spinal cord. It is observed to contain cells and cylindrical tubes; these are formed of an envelope of areolar and fibrous tissue and of a semi-liquid medullary substance (myelin of Virchow), which refracts light strongly, and is sometimes observed to flow out when a nerve is cut. These tubes are united in bundles which are enveloped in a colourless, lustrous, and fibrous membrane sometimes called *neurilemma*.

This membrane may be rendered apparent by treating nervous tissue with a cold dilute solution of caustic potassa, which dissolves the nervous substance with the exception of the neurilemma. This membrane is dissolved, on the contrary, by hydrochloric acid and strong sulphuric acid; it is not coloured yellow by nitric acids.

The ganglions of the nervous centres are formed of cells of variable size, composed of a very thin envelope and a nucleus containing a dense liquid, in which are granules in suspension.

Concentrated alkaline solutions attack and **dissolve**

the nerve cells and tubes. Strong inorganic acids shorten and thicken the fibres. An aqueous solution of iodine colours them bright yellow. A mixture of mercurous and mercuric nitrates renders them rigid and tenacious.

The reaction of the nerves appears to be neutral during life, it becomes acid after death, and finally, at the moment when putrefaction sets in, it has an alkaline reaction.

Different investigations made recently on the matter of the nerves and brain have shown that we are far from completely understanding its compositions; Liebrich's *protagon* is now regarded as a mixture of W. Müller's cerebrin and lecithin, and the same may be said of Köhler's myeloidin and myeloidinic acid.

THE CONSTITUENTS OF THE BRAIN, more or less constant and normal, thus far determined with apparent certainty are:—*Water*,—albuminoid bodies resembling *myosin*,—*elastin* (?)—*neurokeratin*,—*nuclein*,—*collagen*,—soluble *albumen*, coagulating at 75°,—*cerebrin* and *lecithin*,—*glycerinphosphoric acid*,—*fats* (?),—*cholesterin*, *inosit*,—*hypoxanthin*, *xanthin*, *kreatin*,—*lactates*,—*volatile fatty acids* and *uric acid*. Inorganic substances *calcium*, *potassium* and *magnesium phosphates*, *iron oxide*, *silica*, *alkaline sulphates*, *sodium chloride*, and *fluorine*. (Horsford.)

Although very extended and repeated investigations of the chemical nature of the brain have been made,

CONSTITUENTS OF THE BRAIN. 395

it yet remains, chemically, one of the most incompletely-known animal organs.

It was W. Müller who first obtained the nitrogenous neutral body from the brain called *cerebrin*. It is extracted on coagulating by heat an aqueous extract of the cerebral substance; this coagulum is separated and washed with water, and treated while boiling with a mixture of alcohol and ether, and filtered hot. White flakes separate out of the solution, which contains cholesterin, lecithin, and cerebrin. This matter is the cerebric acid of Frémy. Cerebrin has the formula $C_{17}H_{33}NO_3$.

It is dissolved by sulphuric acid, the solution being of a purple colour. It is rendered resinous by hydrochloric acid, and is transformed by boiling nitric acid into an oil which solidifies on cooling. Gobley claims to have also found cerebrin in the yolk of eggs.

E. Bourgoin (60–[2] 21 482) purifies cerebrin from the phosphorous compound which ordinarily adheres to it by treating the same with a sufficient quantity of strong alcohol, and gradually warming; the cerebrin dissolves before the alcohol begins to boil, and the phosphorous compound deposits itself on the bottom of the vessel; the cerebrin separates out of the decanted solution on cooling. The cerebrin should be again subjected to a similar treatment. Bourgoin regards the *protagon* of Liebreich (36–1865–647) as a mixture of cerebrin with this phosphorous compound.

Pure cerebrin shows the following composition :—

Carbon	66.35
Hydrogen	10.96
Nitrogen	2.29
Oxygen	20.40

Lecithin, though a constituent of the brain, is best obtained from the yolk of eggs. It is an imperfectly crystallizable body, easily fusible, with a waxy lustre, soluble in ether and alcohol, and in general very easily decomposed. There appear to be various lecithines with different radicles; one of the most common appears to have the radicle of stearic acid, its empirical formula being $C_{44}H_{90}NPO_8$. (Thudicum.)

The mineral salts constitute about 5 per cent. by weight of the brain in a dry state. When the brain is in full action the elimination of phosphorus appears to be greater than when it is in repose, since the quantity of alkaline phosphates in the urine increases.

The composition of the spinal cord, of the medulla oblongata, of the nervous fibres and ganglions is very similar to that of the cerebral substance. The medulla oblongata contains the largest proportion of fatty bodies.

OSSEOUS TISSUE.—The bones are formed of solid mineral matter (about 70 per cent.), and of an organic cartilaginous tissue, in which is found a principle called *ossein*, furnishing gelatin with boiling water. The bony structure is pierced with numerous cavities. Many are visible to the naked eye; others are extremely minute canals, which penetrate in all directions, forming a complete network, which admits of communication

CONSTITUENTS OF BONE. 397

between the most remote points of the structure; these canals are concerned in the nutrition of the tissue.

The medullary cavity and the cells of spongy bones have a membranous cellular tissue and blood-vessels.

The canaliculi contain only nerves and blood-vessels.

Marrow is formed, according to Berzelius, of—

Fat	96
Blood-vessels, membrane	1
Extractive substances	3
	100

According to Eylerch, the fatty matter of marrow is constituted of three ethers of **glycerin,** whose acids are the palmitic, medullic, and elaidic.

The first is the most abundant.

Bones deprived of their fat and periosteum, are, according to Berzelius, composed of—

		Man.
Mineral portion	Calcium phosphate (tribasic)	53.04
	Calcium carbonate	11.30
	Magnesium phosphate	1.16
	Sodium chloride and carbonate	1.20
Organic portion	Cartilage (ossein)	32.17
	Blood-vessels	1.13
		100.00

Ossein has, as a special characteristic, the property of being transformed by the action of boiling water into gelatin. The membrane which covers the walls of the osseous canals is formed of an albuminoid substance insoluble in boiling water. Nitrogenous bodies derived from the blood-vessels and nerves are also found in the bones, as well as fatty matters.

On treating bones with a dilute alkaline solution the ossein is dissolved, and the mineral portion remains, retaining its original shape.

The composition of bone does not vary greatly with age, except that the hard and compact portion of the bones diminishes in aged people, and is replaced by a spongy and more brittle material; also in children it contains more water, and is more elastic. It has been observed that in lower animals the proportion of calcium carbonate increases with age.

The composition of the bones of different species of animals differs but little; yet the bones of birds and herbivorous mammalia are richer in calcium salts than the bones of carnivora and reptiles.

The bones of the limbs contain more inorganic mineral matter than those of the trunk; the humerus and femur contain more than the other long bones; these also contain less fatty matters than the short bones; the flat bones contain the largest proportion of water.

According to Fremy, the bones are not formed by an incrustation of the mineral portion in the cartilaginous tissue, as is generally believed, but by a juxtaposition of osseous matter, particle by particle; for the rudimentary parts of the bones of the fœtus have the same composition as the bones of full-grown persons, and the composition of the bones does not essentially vary with age.

C. Aeby (18-[2] 10-408) also is of the opinion that the cartilage and calcium phosphate of the bones are not combined, and that the organic foundation of the bones simply induces ossification without entering into chemical relations with the calcium phosphate.

Bones are used in the arts for the manufacture of animal charcoal or bone black, phosphorus, and gelatin. Grease is likewise extracted from them. They also serve as material for a great variety of useful and fancy articles.

GELATIN.—The organic portion of bones is separated from the mineral portions on treatment with dilute hydrogen chloride. The salts are thereby dissolved; this organic substance alone remains, and, while retaining the form of the bone, is flexible, yellowish, and translucent. This substance, formed almost exclusively of ossein, becomes hard on drying, and again pliable and elastic when placed in water for a short time. Submitted to the action of boiling water, it is transformed into gelatin. In making gelatin bones are first treated with boiling water. The grease is thereby separated out and removed. The bones are then placed in a digester with water, and submitted to a pressure of several atmospheres; the gelatin is almost completely dissolved, and the mineral portion remains insoluble. These degelatinized bones form an excellent manure.

The transformation of ossein is more rapid with the bones of a young animal than those of an adult.

The ossein is not combined with the calcium, as can be very easily proven, for, if a few grammes of bones and a quantity of ossein equal in weight to that which exists in these bones be treated with boiling water the transformation is as rapid in one case as in the other, during the first part of the process.

The proportion of gelatin produced by the bone then diminishes; but this is due to the fact that the calcium salts of the exterior layers protect the interior portions from the action of the boiling water; but if the surface of the bone be scraped the action of the boiling water again commences.

Pure gelatin, $C_6H_{10}N_2O_2$ (?), when dry is colourless, or very slightly yellow; elastic and insoluble in alcohol and ether. It swells in cold water, and dissolves in boiling water. It turns the plane of polarization to the left. The solution, on cooling, changes into a gelatinous mass. provided it has not been boiled too long with water. One per cent. of gelatin is sufficient to form a jelly; and sulphuric acid converts it into glycocol. It forms with tannin an insoluble and imputrescible compound, and this chemical action is the basis of the art of tanning. C. Voit (11-8-297) has shown that gelatin is capable of partially replacing albumen and fat, as a food.

Pathological States.—In arthritis, or gout, the articulations become encrusted with concretions, called *arthritic calculi.*

ARTHRITIC CONCRETIONS.

Water	10.3
Animal matter	19.5
Uric acid	20.0
Sodium hydrate	20.0
Lime	10.0
Potassium chloride	2.2
Sodium chloride	18.0
	100.0

(Sebastian).

Exostosis is an affection in which osseous tumours are developed on the bone. An analysis gave:—

	Exostosis.	The bone in the vicinity.
Organic substance	46.0	41.6
Calcium phosphate	30.0	41.6
„ carbonate	14.0	8.2
Soluble salts	10.0	8.6
	100.0	100.0

In *caries* of bone the inorganic portion of the bone is destroyed, while the organic portion remains almost intact. We owe to Von Bibra the following analyses of cases of caries:—

	Tibia at the point amputated.	Tibia taken 6 centimetres from the joint.	Portion of the Astragalus taken from the centre of the caries.
Inorganic substances	61.80	42.10	18.54
Organic ,,	38.20	57.90	81.46
	100.00	100.00	100.00

In *rachitis* the mineral portion is removed to such an extent that the bones become incapable of supporting the body. The ossein is also changed, since boiling water no longer furnishes gelatin with these bones.

Bones of a rachitic child, analyzed by Marchand, contained:—

	Vertebra.	Femur.	Radius.	Sternum.
Cartilages	75.22	72.20	71.25	61.20
Fat	6.12	7.20	7.50	9.34
Calcium phosphates	12.56	14.78	15.11	21.35
Magnesium phosphates	0.92	0.80	0.78	0.72
Calcium carbonate	3.20	3.00	3.15	3.70
Calcium sulphate, Sodium sulphate	0.98	1.02	1.00	1.68
Sodium chloride, calcium fluoride, iron, etc.	1.00	1.00	1.20	2.01
	100.00	100.00	100.00	100.00

Osseous tissues gradually decompose after death. In time nothing remains but the mineral portions, yet this action is very slow, as organic matter has been found in bones buried for several centuries. The character of the soil or other medium in which bones are placed has a great influence upon the rapidity of this change.

The ossein which has not yet been wholly decomposed has the same characters as ossein from fresh bones; it is capable of furnishing gelatin.

DENTAL TISSUES.—Three substances are distinguished in the teeth: the *dentine*, which forms the greater part of the teeth; the *cement*, which covers the cervix and roots; and the *enamel*.

The cement has a structure, similar to that of the bones. It has a cavity which contains the nerves and blood-vessels, and in which arise the little canals which ramify and penetrate to the surface of the teeth. Treated with an acid, it parts with its inorganic constituents, and there remains an organic residue capable of furnishing gelatin, according to some authors, though denied by Hoppe-Seyler. The cement has the composition, substantially, of the bones.

The enamel is hard and brittle; it contains about 90 per cent. of calcium phosphate, and a considerable quanity of calcium fluoride, and only 2 to 6 per cent. of organic substances. When treated with dilute hydrogen chloride, the calcium phosphate dissolves, and prismatic fibres remain, which are not attacked by boiling water, and comport themselves like epithelium.

Berzelius found in the teeth:—

Organic matter	28.0
Calcium phosphate	64.4
Magnesium phosphate	1.0
Calcium carbonate	5.3
Sodium ,, and chloride	1.3
Water, animal matter, alkali (traces)	0.0
	100.0

The inorganic portion, according to Frémy, consists of:—

	Ash.	Calcium Phosphate.	Magnesium Phosphate.	Calcium Carbonate.
Dentine	76.8	70.3	4.3	2.2
Cement	67.1	60.7	1.2	2.9
Enamel	96.9	90.5	traces	2.2

Minute amounts of chlorine and fluorine exist especially in the enamel.

The following are more recent analyses by Aeby:—

	Cement.	Dentine.
Calcium phosphate	61.32	93.35
,, oxide	5.27	0.86
,, carbonate	1.61	4.80
,, sulphate	0.09	0.12
Magnesium carbonate	0.75	0.78
Ferric oxide	0.10	0.09
Organic substances	27.70	3.60

Molar teeth appear to contain more mineral matter than incisors (Bibra). The relation of the calcium phosphate to the calcium combined with carbonic acid, and in some analyses with chlorine and fluorine, suggests an analogy between the composition of the enamel and the mineral apátite.

CHEMISTRY OF THE EYE

The sclerotic coat dissolves almost completely in boiling water, and the liquid obtained is a solution of gelatin and chondrin.

The cornea furnishes chondrin with boiling water; it also contains myosin and an alkaline albuminate.

The choroid coat, on being boiled with water, also furnishes gelatin.

The following analysis of the crystalline humour was made by Berzelius :—

Water	58.0
Albuminous matter	35.9
Aqueous extract and salts	2.4
Alcoholic extract	1.3
Membrane	2.4
	100.0

The albuminous matter coagulates in certain cases, and cataract is then produced, on account of the opacity of the crystalline lens.

Lassaigne analyzed the opaque crystalline lens of the eye of a horse, and found—

Coagulated albuminous matter	29.3
Calcium phosphate	51.4
„ carbonate	1.6
Portion soluble in water	17.7
	100.0

The iris is chiefly elastin and connective tissue.

The retina is an expansion of the optic nerve, which has the composition—

Water	92.90
Albumen	6.25
Fatty substances	.85
	100.00

AQUEOUS HUMOUR OF THE EYE.—Berzelius found in this liquid—

Water	98.10
Lactate, chloride of sodium	1.15
Sodium hydrate	0.75
	100.00

It also contains a small quantity of albumen.

EXUDATIONS.

The name *exudations* is given to liquids formed at the expense of the blood, in consequence of an inflammation which arrests the circulation of this fluid.

Exudations differ from transudations by containing fibrin, much albumen and blood globules, and in being more dense.

PUS

Is a yellowish-white, viscous, neutral liquid, or alkaline if the pus is unhealthy.

It is formed, like the blood, of a liquid (*serum*) in which are corpuscles. These are about .01 mm. in diameter; they contain a viscous liquid and nuclei enclosed in a membrane. The colourless globules of mucus and lymph resemble these corpuscles; they are designated in general as *cystoid* corpuscles.

Pus exposed to the air usually becomes acid, producing margaric, butyric, and other homologous acids. Ammonium sulphide is afterwards formed, and the mass undergoes putrid fermentation.

Pus contains 15 to 16 per cent. of soluble matter, the most important of which is albumen. The existence of a substance called *pyin* has been detected in it, but according to Lehman this body is an abnormal product. It generally contains a larger proportion of soluble salts than the serum of the blood.

Boedecker found in a pus slightly alkaline:—

Water	88.76
Albumen	4.38
Pyin	4.65
Fatty bodies and cholesterin	1.09
Sodium chloride	0.59
Other alkaline salts	0.32
Earthy phosphates	0.21
	100.00

Certain varieties of pus have the property of imparting a blue tinge to linen. Fordos has discovered the principle which produces this coloration: it is a crystalline substance which he has named *pyocyanin*.

Pus swells, and assumes the appearance of gelatin on being mixed with ammonium hydrate. This reaction distinguishes it from mucus.

Pure pus, placed in a vessel and allowed to remain for several hours, separates into two layers. The lower, curdy layer contains the globules and the solids; the upper, opalescent layer constitutes the serum.

C. Robin gives the following analysis of the serum in 1,000 parts:—

Water	937.86 to	970.55
Sodium phosphate	3.11 ,,	4.66
Phosphate of soda	traces ,,	2.22
Earthy and ammonio-magnesium phosphates	0.50 ,,	2.20
Sulphates and carbonates of sodium and potassium	1.87 ,,	3.11
Salts of iron and silica	.16 ,,	.96
Salts with organic acids, formiates, butyrates, valeriates, etc.	traces ,,	1.00
Leucin, tyrosin, and extractive substances	15.00 ,,	20.00
Serolin	1.00 ,,	8.30
Cholesterin	3.50 ,,	10.00
Fatty bodies	10.00 ,,	19.00
Lecithin	6.00 ,,	10.00
Meta-albumen and serin	11.00 ,,	48.00

Among the extractive substances there have been found: Paraglobulin, tyrosin, leucin, xanthin, urea, glucose (in diabetes), bilirubin, uric and chlorrhodinic acids (in necrosis), and a special pus product, hydropsin.

LIST OF ORIGINAL AUTHORITIES.

1. Annalen der Chemie und Pharmacie; v. Liebig u. Wöhler.
2. Annalen der Physik und Chemie von Poggendorf.
3. Archiv der Pharmacie.
4. Bulletin de la société d'encourag.
5. Bulletin de la société de Mulhouse.
6. The Engineer.
7. Chemisches Centralblatt.
8. Chemical News.
9. Comptes rendus.
10. Deutsche Industriezeit.
11. Zeitschrift für Biologie.
12. Gewerbeblatt, Sächsisches
13. ,, Breslauer.
14. ,, Hessisches.
15. ,, Würtemberger.
16. Wieck's Illustr. deutsch. Gewerbztg.
17. Journal de Pharmacie et de Chimie.
18. Journal für praktische Chemie.
19. Bayr. Industrie u. Gewerbeblatt.
20. London Journal of Arts.
21. Lehrbuch der physiol. Chemie. Gorup-Besan. Fourth Ed. 1878.
22. Mittheilungen des Gewerbevereins für Hannover.
23. Reimann's Färberzeitung.
24. Pharmaceut. Centralhalle v. Hager.
25. Photogr. Archiv. v. Liesegang.
26. Polytechn. Centralblatt.
27. Mechanics' Magazine.
28. Dingler's Polytechn. Journal.
29. Polytechn. Notizblatt v. Böttger.
30. Milchzeitung (Dantzic).
31. Practical Mechanics' Journal.
32. Quarterly Journ. of the Chem. Soc.

LIST OF ORIGINAL AUTHORITIES.

33. Ackermann'sGewerbezeitung
34. Repertory of patent inventions.
35. Technologiste.
36. Jahresbericht der Chemie
37. Zeitschrift für analytische Chemie.
38. Journal of Applied Chemistry.
39. Zeitschrift des allgem. österreich. Apotheker-Vereins.
40. Pharmaceut. Zeitschr. f. Russland.
41. Wien. Acad. Ber.
42. Neues Jahrbuch für Pharmacie.
43. Berg- und hüttenmänn. Zeitung.
44. The Lancet (London).
45. Der Bierbrauer (Leipsic).
46. Archiv. Pharm.
47. Gazetta Chimica Italiana
48. Elsner's Chem. -techn. Mittheilgn.
49. Industrieblätter v. Hager und Jacobsen.
50. Photographische Mittheilungen v. H. Vogel.
51. Zeitschrift des Vereins für die Rübenzuckerindustrie
52. American Jour. of Pharmacy.
53. Photographische Correspondenz v. Hornig.
54. Bulletin belge de la photographie par Deltenre-Walker.
55. **London Royal Society Proceedings.**
56. Chemisch-Technisch Repertorium.
57. Neue Deut. Gewb.-Zeitg.
58. Wagner's Jahresbericht der chem. Technologie.
59. Würzburg. gemeinn. Wochenschr.
60. Berichte der deutschen chem. Gesellschaft.
61. Proceedings of the French Association for the Advancement of Science.
62. Lyon Medicale.
63. Scientific American.
64. American Artizan.
65. Journal für Gasbeleuchtung.
66. Moniteur Scientifique
67. Badische Gewerbezeitung.
68. Der Naturforscher (Berlin).
69. Deutsche Weinzeitung.
70. Annales du Génie civil.
71. Les Mondes.
72. Annales de Chimie et de Physique.
73. Deutsche Gerberzeitung.
74. Chicago Pharmacist.
75. Neues Repert. der Pharm.
76. Nature (London).
77. Nacquet's Modern Chemistry.
78. Schweizer. Zeitschr. f. Pharmacie.
79. Virchow's Archiv.
80. American Journal of Science.
81. Zeitschrift f. d. gesammten Naturwissenschaften.
82. Zeitschrift für Chemie.
83. **Photographic News.**

84. Brit. Journ. of Photography.
85. New Remedies.
86. Philadelphia Photographer.
87. London Medical News.
88. Moniteur Industriel.
89. Jahresbericht der Thierchemie.
90. Centralblatt f. d. Papierfabrik.
91. Engineering.
92. Propagation Industrielle.
93. Journal de l'Agriculture p. Barral (Paris).
94. Proceedings of the Am. Pharm. Ass'n.
95. Revista Pharmaceutica (Buenos Ayres).
96. Journal for Pharmaci (Copenhagen).
97. Bulletin de la Société Chimique (Paris).
98. Popular Science Monthly.
99. Journ. of the Franklin Institute.
100. American Chemist.
101. Kunst und Gewerbe (Nuremberg).
102. Neues Handwoerterbuch der Chemie.
103. Jacobsen's Chem.-tech. Repertorium.
104. Philosophical Magazine (London).
105. Pharm. Journal and Transactions.
106. Pharm. Zeitung (Bunzlau).
107. Zeitschrift für Chem. Grossgewerbe.
108. Die Chem. Industrie auf der Austellung in Philadelphia.
109. Zeitschrift für Physiolog. Chemie. Hoppe-Seyler.
110. Moniteur de la teinture.

INDEX.

Acenapthene, $C_{12}H_{10}=154$.. 38
Acetamide, $C_2 H_5 NO=59$.. 136
Acetanilide, $C_8 H_9 NO=135$. 130
Acetic oxide $C_4 H_6 O_3 =102$ 103
Acetochlorhydric glycol...... 63
Acetone, $C_3 H_6 O=58$....99, 108
Acetyl acetate, $C_4 N_9 O_3$ ... 103
Acetyl chloride, $C_2 ClH_3 O$. 103
Acetyl hydride or aldehyd,
  $C_2 H_4 O=44$............. 86
Acetylamine, $C_2 H_5 N=43$.. 129
Acetylene, $C_2 H_2 =26$...... 18
Acetylide, cuprous......... 19
Acid, acetic, $C_2 H_4 O_5 =60$.. 99
Acid, aconitic, $C_6 H_6 O_6 =95$ 174
Acid, acrylic, $C_3 H_4 O_2 =72$. 91
Acid, adipic, $C_6 H_{10}O_4 =148$ 91
Acid, alloxanic, $C_4 H_4 N_2 O_5$ 125
Acid, alpha-cymic, $C_{11}H_{14}O_2$ 91
Acid, amalic, $C_6 H_7 N_2 O_4$ .. 169
Acid, anchoic, $C_9 H_{16}O_4 =188$ 93
Acid, angelic, $C_5 H_8 O_2 =108$ 91
Acid, anisic, $C_8 H_8 O_3 =152$. 92
Acid, arabic, $C_6 H_{10}O_5 =342$ 217
Acid, arichidic, $C_{20}H_{40}O_2$ .. 90
Acid, atropic, $C_9 H_8 O_2 =148$ 164
Acid, benzoic, $C_7 H_6 O_2 =126$
  ....91, 109, 126
Acid, benzoglycolic 126
Acid, butyric, $C_4 H_8 O_2$ ...90, 108
Acid, caffetannic.......... 196

Acid, camphic, $C_{10}H_{16}O_{20}=91$ 168
Acid, campholic, $C_{19}H_{18}O_4$ .. 91
Acid, camphoric, $C_{10}H_{18}O_4$ 41, 93
Acid, caprylic, $C_8 H_{16}O_2$ ... 90
Acid, caproic, $C_6 H_{12}O_2 =116$ 90
Acid, capric, $C_{10}H_{20}O_2 =172$ 90
Acid, carballylic, $C_6 H_8 O_6$. 95
Acid, carbamic, $CH_3 NO_2$ ... 11
Acid, carbazotic, (Picric)
  $CH_3 N_3 O_7 =229$......... 33
Acid, carbolic, $C_6 H_6 O=94$. 32
Acid, carbonic, $C_2 H_3 O=62$. 92
Acid, catechic............. 196
Acid, cerotic, $C_{27}H_{54}O$ ...90, 180
Acid, chelidonic, $C_7 H_4 O_6$ .. 95
Acid, chlorbenzoic, $C_7 H_5 ClO$
  $=130.5$.... 160
Acid, cholalic, $C_{24}H_{40}O_5 =408$ 95
Acid, cholesteric, $C_8 H_{10}O_5$ .. 95
Acid, choloidic, $C_{24}H_{38}O_4 =390$ 94
Acid, cinnamic, $C_9 H_8 O_2 =$
  148.................91, 111
Acid, citraconic, $C_5 H_6 O_4$ 93, 121
Acid, citric, $C_6 H_8 O_7, H_2 O=$
  $192+18$................120, 95
Acid, coccinic, $C_{18}H_{26}O_2$ ... 90
Acid, comenic, $C_6 H_4 O_5$ ... 95
Acid, coumaric, $C_9 H_8 O_3$ .. 93
Acid, croconic, $C_5 H_2 O_5$ .. 95
Acid, crotonic, $C_4 H_6 O_2$ ..91, 178
Acid, cumic, $C_{10}H_{12}O_2 =164$ 91

INDEX.

Acid, cyanacetic,
$C_2 H_3 (CN) O_2 = 85$ 103
Acid, cyanhydric, $HCN = 27$. 161
Acid, dextroracemic 117
Acid, dialuric, $C_4 H_4 N_2 O_4$ 125
Acid, dinitrobenzoic,
$C_7 H_4 (NO_2)_2 O_2 = 212$ ... 110
Acid, doeglic, $C_{19}H_{36}O_2 = 296$ 91
Acid, elaidic 177
Acid, erucic, $C_{22}H_{42}O_2 = 338$. 91
Acid, ethalic, $C_{16}H_{32}O_2 = 256$ 179
Acid, ethylsulphuric,
$C_2 H_5 HSO_4 = 126$ 71
Acid, formic, $CH_2 O_2 = 50.98$, 90
Acid, fumaric, $C_4 H_4 O_4 = 116$ 93
Acid, gallic, $C_7 H_6 O_5$ ..95, 197
Acid, glucic, $C_{12} H_8 O_9 = 306$ 186
Acid, glyceric, $C_3 H_6 O_4$ ... 93
Acid, glycolic, $C_2 H_4 O_3$ .60, 92
Acid, guaiacic, $C_6 H_8 O_3$ ... 92
Acid, gummic, $C_{12} H_{22} O_{11}$.. 217
Acid, hippuric, $C_9 H_9 NO_3$ .. 125
Acid, insolinic, $C_9 H_8 O_4$ ... 94
Acid, itaconic, $C_5 H_6 O_4$ ... 121
Acid, lactic, $C_3 H_6 O_3$ ..92, 122
Acid, lauric, $C_{12}H_{24}O_2 = 200$ 90
Acid, leucic, $C_6 H_{12}O_3 = 132$. 92
Acid, lichenstearic, $C_9 H_{14}O_3$ 92
Acid, lithic, $C_5 H_4 N_4 O_3$ .. 123
Acid, lithofellic, $C_{20}I_{36}O_4$ .. 93
Acid, malic, $C_4 H_6 O_5 = 134$ 115
Acid, malonic, $C_3 H_4 O_4$ ... 93
Acid, mannitic 183
Acid, margaric, $C_{17}H_{34}O_2$.. 177
Acid, meconic, $C_7 H_4 O$ 143
Acid, melissic, $C_{30}H_{60}O_2$ .. 90

Acid, mellitic, $C_4 H_2 O_4$ 94
Acid, mesoxalic, $C_3 H_2 O_5$ .. 94
Acid, metagummic 217
Acid, monochloracetic,
$C_2 Cl H_3 O_2 = 94.5$ 201
Acid, moringic, $C_{15}H_{28}O_2$ .. 91
Acid, morintannic 196
Acid, mucic, $C_6 H_5 O_8 = 205$ 95
Acid, myristic, $C_{14}H_{28}O_{20}$... 90
Acid, œnanthalic, $C_7 H_{14}O_2$ 90
Acid, œnanthic, $C_{14}H_{28}O_3$ .. 92
Acid, oleic, $C_{18}H_{34}O_2 = 282$. 91
Acid, opianic 127
Acid, oxalic, $C_2 H_2 O_4$ ..93, 112
Acid, oxamic, $C_2 H_3 NO_3$ .. 11
Acid, oxybenzoic, $C_7 H_6 O_3$ 195
Acid, oxybutyric, $C_4 H_8 O_3$ 92
Acid, oxycuminic, $C_{10}H_{12}O_3$ 92
Acid, oxynapthalic, $C_{10}H_6 O_4$ 94
Acid, oxyvaleric, $C_5 H_{10}O_3$ .. 92
Acid, palmitic, $C_{16}H_{32}O_2$ .90, 177
Acid, parabanic, $C_3 H_2 N_2 O_3$ 125
Acid, parafinic, $C_{21}H_{48}O_2$ .. 23
Acid, paralactic 122
Acid, paramalic, $C_4 H_4 O_4$ .. 116
Acid, paratartaric 117
Acid, pectic, $C_{16}H_{22}O_5 = 294$. 218
Acid, pectosic 218
Acid, pelargonic, $C_9 H_{18}O_2$ ... 90
Acid, phenic, $C_6 H_6 O = 94$.. 32
Acid, phenylsulphuric,
$C_6 H_6 O_4 S = 174$ 32
Acid, phloretic, $C_9 H_{10}O_3$ .. 92
Acid, phtalic, $C_8 H_6 O_4 = 150$ 94
Acid, physetoric, $C_{16}H_{30}O_2$ .. 91
Acid, picric, $C_6 H_3 (NO_2)_3 O$ 33

INDEX. 415

	PAGE.
Acid, pimelic, $C_7 H_{12}O_4$	93
Acid, pinaric, $C_{20}H_{30}O_2 = 302$	41
Acid, pinic, $C_{20}H_{30}O_2 = 302$..	91
Acid, piperic, $C_{12}H_{10}O_4 = 218$	94
Acid, propionic, $C_3 H_6 O_2$ 78,	90
Acid, prussic, $HCN = 27$. ...	161
Acid, pyrogallic. $C_6 H_6 O_3$ ..	198
Acid, pyroligneous.........	100
Acid, pyromeconic, $C_5 H_4 O_3$	92
Acid, pyrotartaric, $C_5 H_8 O_4$	
$=132$..................93,	117
Acid, pyroterebic, $C_6 H_{10}O_2$ ..	91
Acid, pyruvic, $C_3 H_4 O_3 = 88$	92
Acid, quinic, $C_7 H_{12}O_6 = 144$.	93
Acid, quinotannic..........	196
Acid, racemic, $C_4 H_6 O_6 = 150$	117
Acid, ricinoleic, $C_{18}H_{34}O_3$, 92,	180
Acid, roccellic, $C_{17}H_{32}O_4$ ..	93
Acid, salicylic, $C_7 H_5 O_3$ 195,32,92	
Acid, sarcolactic...........	122
Acid, scammonic, $C_{15}H_{28}O_3$	92
Acid, sebic, $C_{10}H_{18}O_4 = 202$..	93
Acid, sorbic, $C_6 H_8 O_2 = 112$.	91
Acid, stearic, $C_{18}H_{36}O_2$ ..90,	177
Acid, suberic, $C_8 H_{14}O_4 = 174$	93
Acid, succinic, $C_4 H_6 O_4$ 93,	115
Acid, sulphocarbolic,	
$C_6 H_6 SO_4 = 174$.........	33
Acid, sulphoglucic..........	185
Acid, sylvic, $C_{20}H_{30}O_2 = 302$.	41
Acid, tannic, $C_{27}H_{22}O_{17} = 618$	196
Acid, tartaric, $C_4 H_6 O_6$ ...116,	95
Acid, tartrelic, $C_4 H_4 O_5$ ...	117
Acid, tartronic, $C_3 H_4 O_5$ ..	94
Acid, terebic. $C_7 H_{10}O_4 = 158$	93
Acid, terechrysic, $C_8 H_6 O_4$	94

	PAGE.
Acid, thionuric,	
$C_4 H_5 NO_3 SO_3 = 195$....	125
Acid, thymotic, $C_{11}H_{14}O_3$ ..	92
Acid, toluic, $C_8 H_8 O_2 = 136$	91
Acid, trichloracetic,	
$HC_2 Cl_3 O_2 = 163.5$.......	102
Acid, tropic, $C_9 H_{10}O_3 = 166$.	164
Acid, uric, $C_5 H_4 N_4 O_3 = 168$	123
Acid, valeric or valerianic,	
$C_6 H_{10}O_2 = 102$........109,	90
Acid, veratric, $C_9 H_{10}O_3$ ...	94
Acid, xylic, $C_9 H_{10}O_2 = 150$.	91
Acids.....................	95
Acids, aromatic............	91
Acids, fatty...............	90
Acids, general methods of	
preparation,.............	96
Acids, organic.............	90
Acids, defined.............	95
Acids, polyatomic..........	112
Acids, pyro................	97
Aconitina, $C_{30}H_{47}NO_7 = 533$.	165
Albumen	228
Alcohol, amylic, $C_5 H_{12}O$. 56,	45
Alcohol, benzyl, $C_7 H_8 O = 108$	
....	46
Alcohol, butyl, $C_4 H_{10}O = 64$	45
Alcohol, ceryl, $C_{27}H_{56}O = 396$	45
Alcohol, cholesteryl.......	46
Alcohol, cinnyl, $C_9 H_{10}O$ ..	46
Alcohol, cuneol............	46
Alcohol, cymol, $C_{10}H_{14}O$ ..	46
Alcohol, melissic, $C_{30}H_{62}O$ ..	180
Alcohol, methyl, $CH_4 O$ ..45,	46
Alcohol, myricyl, $C_{30}H_{62}O$ ..	45
Alcohol, octyl, $C_8 H_{18}O = 130$	45

INDEX.

	PAGE
Alcohol, ordinary, or ethyl, $C_2H_6O=46$	49
Alcohol, propyl, C_3H_8O...	45
Alcohol, sexdecyl, $C_{16}H_{34}O$..	45
Alcohol, sextyl, $C_6H_{14}O$.....	45
Alcohol, vinyl, $C_2H_6O=46$	45
Alcohol, xylyl, $C_8H_{10}O=122$	46
Alcohols, diatomic	58
Alcohols, monatomic	44
Alcohols, polyatomic	59
Alcohols, sulphur	82
Alcohols, selenium	82
Alcohols, tellurium	82
Alcohols, tetratomic	59
Alcohols, triatomic	64
Aldehyds	86
Alizarin, $C_{10}H_6O_3=174$...	39
Alkalamides	136
Alkaloids	127
Allantoin, $C_4H_6N_1O_3=158$	124
Alloxan, $C_4H_4N_2O_5=160$.	125
Alloxantin, $C_8H_{10}N_4O_{10}$..	123
Allyl iodide, $C_3H_5I=168$..	57
Allyl sulphide, $C_6H_{10}S=114$	57
Allyl sulpho-cyanide, $C_4H_5NS=99$	57
Allylamine, $C_3H_7N=57$...	127
Allylene, $C_3H_4=40$	20
Amane, $C_5H_{12}=72$	23
Amber	26, 42
Amides	136
Amidoxypropyl, $C_3H_4(NH_2)O=72$	75
Amines	133
Ammelide	172

	PAGE
Ammonia aldehydate, $C_2H_4ONH_3=61$	87
Ammonia citrate of iron...	121
Ammoniacum	43
Ammonias, compounds	131
Ammonium, cyanate, CH_4N_2	172
Ammoniums	137
Ammoniums, quarternary	136
Amygdalin, $C_{20}H_{27}NO_{11}$....	193
Amyl, acetate, $C_7H_{14}O_3$..	56
Amyl, chloride, $C_5H_{11}Cl$..	56
Amyl, hydride, $C_5H_{12}=72$.	23
Amylamine, $C_5H_{13}N=87$..	121
Amylene, $C_5H_{10}=70$	23
Anhydride, tartaric, $C_4H_4O_5=132$	117
Aniline	30, 127, 131
Anthracene, $C_{14}H_{10}=178$..	29, 39
Arabin $C_{12}H_{22}O_{11}=342$	217
Arbutin $C_{13}H_{16}O_7=284$....	193
Aricina $C_{23}H_{26}N_2O_4=397$..	129
Arnicin	42
Aromatic compounds	89
Arsines	128
Asphalt	26
Assafœtida	43
Atropia $C_{17}H_{23}NO_3=289$.	164,129
Balsams	41
Bases organic,	125
Bases quarternary,	136
Bassorin	218
Belladona	164
Benzene $C_6H_6=78$	27
Benzine	24
Benzoic aldehyd, C_7H_6O..	86
Benzol, $C_6H_6=78$	27

INDEX. 417

	PAGE.		PAGE.
Benzone	119	Campholic alcohol	117
Benzonitrile	110	Camphor, artificial	37
Benzyl chloride	126	Camphor	40
Benzylene	20	Camphor, monochlor	41
Bezoar	267	Camphor, oxy-	41
Bidecane	28	Camphor of Borneo	58
Bidecyl hydride	23	Cantharidin	168
Bilifulvin	257	Candles	176
Bilirubin	257	Cannabin	42
Biliverdin	257	Caoutchouc	36, 43
Bile	250	Caprylamine	127
Bile, action on food	258	Caramel	190
Bitumen	26	Caramelane	190
Biuret	172	Caramelene	190
Blood	272	Caramelin	190
Blood, action of different gases on the	291	Carbo-hydrates, defined	7
		Carbon dioxide	313
Blood, chemical pathology of the	294	Caries	401
		Carbonic ether	74
Blood, coagulation of	276	Cartilagein	392
Blood, gases of the	288	Casein, animal	226, 233
Blood globules	281	Casein, vegetable	219, 234
Blood, iron of the	287	Castor oil	180
Blood, uses of	293	Castorin	42
Bones	399	Cellulose (cellulin)	202
Borneol	58	Cerasin	217
Brain constituents	394	Cerebrin	395
Brandy	52	Cetene	23
Brucia	161, 129	Chitin	184
Butane	23	Chloral	87
Butter	179	Chloral hydrate	88
Butyl hydride	23	Chloroform	47
Butylamine	128	Chloropropyl	15
Butylene	20, 22	Cholera	296
Cacodyl	79, 105	Cholesterilene	255
Caffeia (caffein)	130, 168	Cholesterin	255

	PAGE		PAGE
Cholesterophan	169	Cyanopropyl	15
Cholin	251	Cyclamin	193
Chondrin	327, 214, 392	Cymene	38
Chondroglucose	392	Cymogene	24
Chyle	269	Cymol	41
Chyme	268	Cystin	353
Chymosine	247	Daphnin	193
Cinchonia (cinchonine)	129, 156	Daturia, (atropia)	164, 129
Cinchonicia (cinchonicine)	158	Decane	24
Cinchonidia (cinchonidine)	158, 129	Dextrin	212, 214
Cinnamene	38	Dental tissue	403
Coagulum	281	Diabetes	327, 347
Codeia	146, 129	Diastase	212
Colchinia	163	Diethylamine	128
Colloidin	375	Diethylpropyl	15
Collodion	208	Diethylenic diamine	170
Colophony	41	Digestion	237
Compound ammonias	131	Digitalin	166
Conia (conine)	141, 129	Digitin	166
Conicin	129	Dimethylphosphine	128
Coniferin	193	Draconyl	38
Convolvulin	193	Dropsy	297
Conylia	141, 129	Dulcite, (dulcose)	183, 181
Cotarnin	147	Duodecylene	23
Cream of Tartar	116	Dysentery	266
Creatin	188, 386	Dystisin	253
Creatinin	386	Elaidin	175
Creosote	34	Elaine	175
Cresofol	29, 34	Elastin	389
Crotonylene	20	Elemi	43
Cumene	28	Emetia	167
Cumidin	127	Emetics	119
Cuprous acetylide	20	Emydin	226
Curari	163	Ergotin	42
Curarina	162	Erythrite	49
		Esculin	193

INDEX.

	PAGE		PAGE
Essence of mirbane	29	Ethyl chloride	75
Essence of thyme	34	Ethyl cyanide	77
Essential oil of cloves	37	Ethyl formiate	9
Essl. oil of bergamot	37	Ethylglycol	61
Essl. oil of copaiba	37	Ethyl-hexyl ether	84
Essl. oil of cubebs	37	Ethyl hydride	23
Essl. oil of elemi	37	Ethyl iodide	76
Essl. oil of juniper	37	Ethyl mercaptan	83
Essl. oil of lemon	37	Ethylmethylaniline	30
Essl. oil of orange	37	Ethyl oxide	69
Essl. oil of pepper	37	Ethyl sulphide	83
Ethal	179	Ethylamine	132, 127
Ethane	13, 15, 23	Ethylene	21
Ethene	13, 15	Ethylene bromide	61
Ether, acetic	73	Ethylene chloride	76
Ether, butyric	81	Ethylene oxide	62
Ether, chlorhydric	75	Eucalin	182
Ether, common	70	Eye, chemistry of the	405
Ether, cyanhydric	77	Excrements	265
Ether, ethyl	70	Excretin	265
Ether, formic	81	Extosis	401
Ether, hydriodic	76	Exudations	407
Ether, hydrosulphuric	83	Fats	174
Ether, œnanthylic	81	Fatty acid series	90
Ether, oxalic	74	Ferment, bile	258
Ether, oxamic	117	Fermentation, acetic	100
Ether, sulphuric	70	Fermentation, alcoholic	49, 181
Ether, valerianic	81	Fermentation, gallic	197
Ether, vinic	70	Fermentation, lactic	122
Ethers	69	Ferrocyanide of potassium	172
Ethers, simple	69	Fibrin	226, 231
Ethers, compound	73	Flesh	382
Ethers, miscellaneous	81	Flour	215
Ethers, mixed	38	Food, respiratory	223
Ethine	13	Food, plastic	224
Ethyl	15	Food, transformation of	321

INDEX.

	PAGE.		PAGE.
Formene	23	Hæmoglobin	284, 226
Frankincense	43	Helicin	194
Fulminates	54	Heptyl hydride	23
Fusel or fousel oil	56	Heptane	23, 24
Galactose	187, 182	Heptylene	22
Gas, illuminating	21	Hexadecane	24
Gasolene	24	Hexadecyl hydride	24
Gastric juice	242	Hexane	23
Gasterase pepsin	247	Hexylene	22
Gelatin	234, 399	Hexyl hydride	23
Glucosane	185	Hoffmann's anodyne	73
Glucose	180, 182, 184, 343	Homologous series	12
Glucose in the liver	323	Honey	192
Glucosides	192, 184	Hydrides	23
Glue	235	Hydrocarbons	18
Gluten	216	Hydrocarbides	18
Glycerin	64	Hydrocarbides, extra-terres-	
Glycocol, zincic	126	trial	40
Glycogen	214, 250, 324	Hydrocephalus fluid	374
Glycol, amyl	59	Hydrogen carbides	18
Glycol, butyl	59	Hydropical fluid	375
Glycol, diethyl	61	Hydrosulphuric Ether	83
Glycol, ethyl	61	Hyosciamine	164
Glycol, hexyl	59	Ictithin	226
Glycol, monochlorhydric	62	Indican	342
Glycol, octyl	59	Indigogen	343
Glycol, ordinary	59	Indiglucin	343
Glycol, propyl	123	Indigo	130
Grape sugar	182	Inosite (inosin)	182
Guano	124	Intestinal concretions	267
Gum	216	Intestinal fluids	264
Gum arabic	217	Intestinal gases	265
Gum resins	41	Inulin	214
Gun-cotton	207	Iodomorphia	145
Hæmatin	286	Iron of the blood	287
Hæmatocrystallin	225	Isatin	38

	PAGE		PAGE
Isologous series	12	Methyl hydride	23
Isomerism	8	Methylamine	131
Jalapin	193	Methylethylamine	128
Jervia	163	Methylphosphine	128
Kerosene	24	Methylpropyl	15
Ketones	40	Milk	376
Lactide	123	Molasses	189
Lactose or lactin	191, 182	Monamines	133
Leather	197	Monochlorcamphor	41
Legamin	219	Monochlorhydrin	66
Leucocythæmia	296	Morphia (Morphine)	143, 129
Levulosan	190	Mucin	227
Levulose	187, 182	Mucus	372
Lichenin	214	Murexide	125
Lymph	270	Muscular power	316
Madder	39	Muscular tissue	283
Maltose	182	Musculin	232
Mannitane	183	Myosin	232
Mannite	181, 183	Mycose	182
Marsh-gas	23	Naphtha	24
Meconin	143, 147	Naphthalamine	128
Melampyrite	181	Naphthalin	27, 38
Melanin	389	Narceia	148, 129
Melezitose	182	Narcotina	129
Melitose	182	Neocytes	337
Mercaptans	82	Neurin	257
Metalbumen	225	Nevrilemma	393
Metamerism	9	Nicotina	139, 129
Metaterebenthene	38	Nicotyl	140
Metastyrol	38	Nicotylia	139, 129
Methane	13, 15, 23	Nitrilebases	124
Methenyl	15	Nitrobenzol	29
Methyl	15	Nitrogenous substances	223
Methyl acetate	9	Nitroglycerine	66
Methyl chloride	47	Nitryls, or cyanhydric ethers	134
Methyl cyanate	131	Nonane	23

	PAGE.
Nonyl hydride	23
Nonylene	22
Nutrition	316
Nutrition, role of mineral compounds in	330
Octane	23
Octylglycol	59
Octyl hydride	23
Octylene	22
Oils, fatty	174
Oils, essential	36
Olein	175
"Oleomargarine"	179
Oleo-resins	42
Opium	142
Orcin	193
Organizable substances	205
Organometallic compounds	78
Ossein	226, 234
Osseous tissues	399
Oxamide	74
Oxanthracene	39
Oxycamphor	41
Oxygen	311
Pancreatic juice	261
Pancreatin	262
Para-arabin	192
Paralbumen	226
Plants, respiration of	201
Plants, nutrition of	204
Polyamines	170
Polymerides	9
Polymerism	9
Populin	193
Pancreatin	262
Pancreatic juice	261

	PAGE.
Paraffin	22, 24
Papaverin	129, 148
Paramorphia	148
Paramylene	22
Parapeptone	249
Pectin	218
Pectose	218
Pentadecane	24
Pentadecyl hydride	24
Pepsin	227, 247
Peptones	225, 249
Petroleum	24
Phenol	32
Phenol, potassic	32
Phenol, trinitric	30
Phenyl	30
Phenyl hydrate	32
Phenylamine	127
Phlorizin	193
Phlorylol	34
Phosphines	128
Phtalidamine	127
Picrotoxin	160
Pinite	181
Piperidine	141
Piperine	141
Pitch, Burgundy	42
Plethora	295
Potassium, formiate	88
Propane	13, 15, 23
Propenyl	15
Propine	13
Propone	13
Propyl	15
Propyl hydride	23
Propylamine	127

INDEX.

	PAGE.		PAGE.
Propylene	22	Rye	216
Proplene iodide	64	Saccharide	186
Protein	225	Saccharoses	182
Ptyalin	212, 227, 238	Salicin	194
Pus	407	Saligenin	194
Pyin	227, 407	Saliva	237
Pyocyanin	408	Saponification	176
Pyrethrin	42	Saponin	193
Pyrocatechin	352	Scurvy	297
Pyrolignite	106	Semen	371
Pyroxylin	207	Serosity	374
Quercite	181	Serum	278
Quercitrin	193	Sinapolin	58
Quinia, (quinine)	151, 129	Sinnamin	58
Quinicia	154, 129	Soaps	176
Quinidia	129	Sodium ethyl	80
Quinidia, oxalate of	155	Sodium sulphocarbolate	33
Quinoidin	158	Solanidia (solanidine)	165
Quinolein, (quinolin)	130,153,157	Solania (Solanine)	165,129,193
Quinovin	193	Sorbin	182
Rachitis	402	Spermaceti	179
Radicles, defined	14	Spirit of Mindererus	105
Radicles, organometallic	78	Stannethyl	79
Radicles, organometalloid	81	Stannethyl iodide	79
Reagent, Fehling's	187	Starch	210
Reagent, Haines'	187	Stearin (stearine)	174
Reagent, Trommer's	186	Stearine candles	176
Resins	25, 41	Stercorin	257, 265
Respiration	272, 301	Stibines	128
Retinasphalt	25	Stibyl	119
Retinite	25	Strychnia (strychnine)	159, 129
Rhigolene	24	Styrol	38
Rice	216	Sucrates	190
Rochelle salt	118	Sugars	181
Rosanilin	31	Sugar of milk	191, 182
Rutylene	20	Sweat	370

INDEX.

	PAGE		PAGE
Synovia	374	Tridecylene	22
Syntonin	229, 232	Triethylamine	135
Tannin	196, 193	Triethylarsine	128
Tartar emetic	116	Triethylenic, diamine	170
Taurin	254	Triethylstibine	128
Teeth	403	Trimethlamine	128
Tetrachloropropyl	15	Trimethylphosphine	128
Tetradecane	24	Tunicin	184, 209
Tetradecyl hydride	24	Turpentine	35
Tetradecylene	22	Types, organic	10
Tetrethylammonium	133	Typhoid fever	266, 296
Thebaia	148, 120	Urinary calculi	353, 368
Theia (theine)	168, 130	Urinary deposits	352, 364
Theobromin	169, 130	Urine	333
Thymol	34	Urine, analysis of	356
Thiosinnamin	58	Urochrome	343
Tissues	388	Uroglaucin	343
Tissues, areolar	388	Urorubrohæmatin	352
Tissues, recticular	387	Urrhodin	343
Tissues, cartilagenous	391	Uroxanthin	343
Tissues, nerve	393	Wax	179
Tobacco	140	Whiskey	52
Toluene	28	Wines	32
Toluidin	127, 130	Wood spirit	49
Transpirations	370	Xylene	28
Trehalose	182	Xylidin	127
Trichlorhydrin	66	Xylyl alcohol	46
Trichloroxypropyl	15	Zinc, ethyl	79
Tridecane	27	Zinc, glycol	79, 126
Triedecyl hydride	24		

www.ingramcontent.com/pod-product-compliance
Lightning Source LLC
Chambersburg PA
CBHW030544300426
44111CB00009B/856